Food and Beverage Connsumption and Health

FISH CONSUMPTION AND HEALTH

FOOD AND BEVERAGE CONSUMPTION AND HEALTH

Handbook of Green Tea and Health Research
Helen McKinley and Mark Jamieson (Editors)
2009. ISBN: 978-1-60741-045-4

Marketing Food to Children and Adolescents
Nicoletta A. Wilks
2009 ISBN: 978-1-60692-913-1

Food Labelling: The FDA's Role in the Selection of Healthy Foods
Ethan C. Lefevre (Editor)
2009. ISBN: 978-1-60692-898-1

Fish Consumption and Health
George P. Gagne and Richard H. Medrano
2009 ISBN: 978- 1-60741-151-2

Food and Beverage Connsumption and Health

FISH CONSUMPTION AND HEALTH

GEORGE P. GAGNE
AND
RICHARD H. MEDRANO
EDITORS

Nova Science Publishers, Inc.
New York

Library of Congress Cataloging-in-Publication Data

Fish consumption and health / editors, George P. Gagne and Richard H. Medrano.
p. ; cm.
Includes bibliographical references and index.
ISBN 978-1-60741-151-2 (hardcover)
1. Seafood--Health aspects. 2. Fish as food. I. Gagne, George P. II. Medrano, Richard H.
[DNLM: 1. Fish Oils--contraindications. 2. Fish Oils--therapeutic use. 3. Fatty Acids, Unsaturated--therapeutic use. 4. Mercury--adverse effects. 5. Seafood--adverse effects. QU 86 F932 2009]
QP144.F56F535 2009
615'.34--dc22

2009001302

Published by Nova Science Publishers, Inc. ✝ New York

Contents

Preface		**vii**
Chapter I	Fish Consumption and Health *Christos Pitsavos, Christina-Maria Kastorini and Christodoulos Stefanadis*	**1**
Chapter II	How to Obtain the Health Benefits Associated with Eating Fish from other Sources *William B. Grant*	**37**
Chapter III	Differences in Understanding of the Risks and Benefits of Fish Consumption in Three Coastal Estuaries in New York and New Jersey *Joanna Burger*	**61**
Chapter IV	Fish Consumption: A Paradox of Good Health *Johnny O. Ogunji*	**77**
Chapter V	Protein-Energy Wasting, Chronic Inflammation and PUFAS: Metabolic and Cardiovascular Risk in ESRD *Gordana Perunicic-Pekovic, Zorica Rasic-Milutinovic, Sladjana Sobajic, Ivana Djuricic and Radojka Maletic*	**93**
Chapter VI	Omega-3 Dietary Supplements and the Risk of Cardiovascular Events: A Systematic Review *Paul E. Marik and Susan Baik*	**105**
Chapter VII	Fish Consumption and Cancer: Summary of Evidence *Faina Linkov, Barbara Stadterman and Emanuela Taioli*	**117**
Chapter VIII	Federal and State Advisories on Mercury in Seafood: An Assessment of Women's Fish Consumption Behaviors and Advisory Awareness *Carole B. Rudra and Michelle A. Williams*	**135**

Chapter IX Emerging Health Risk Due to Shellfish Consumption? **157**
Juan Martínez-Hernández and Victoria Lopez-Rodas and Eduardo Costas

Chapter X Role of Parenteral Fish Oil Lipid Emulsions in Pediatric Nutrition: Benefits and Concerns **167**
Kathleen M. Gura

Chapter XI Do Specific Immunoglobulin E Levels Have any Role in the Diagnostic of Corn Allergy in Children? **177**
Elizabeth Estrada-Reyes, Mirna Toledo-Bahena, and Alejandro A. Nava-Ocampo

Index **181**

Preface

Chapter I - Fish is one of the main components of a healthy diet and an excellent source of numerous beneficial compounds, such as proteins, vitamins, minerals and especially n-3 fatty acids, with several health benefits and particularly as far as cardiovascular disease is concerned. There is good evidence that modest fish consumption of approximately two servings of fish per week lowers the primary risk of coronary artery disease. Fish oil is rich in n-3 fatty acids which have anti-inflammatory, antithrombotic, antiarrhythmic, hypolipidemic, hypotensive and vasodilatory properties. Fish consumption also has beneficial effects in type 2 diabetes, stroke, renal disease, types of cancer and especially breast, endometrial, colorectal, renal and prostate, rheumatoid arthritis, asthma and chronic obstructive pulmonary disease. Furthermore adequate fish consumption is necessary in order to ensure proper brain development and cognitive performance. Deficiencies in n-3 long chain fatty acids or imbalance in the n-6 to n-3 fatty acid ratio have been implicated in a variety of central nervous system disorders, such as Alzheimer's disease and psychiatric illnesses, including depression and mood disorders. On the other hand fish consumption in large amounts has been related to health risks, mainly due to its content in toxic heavy metals, polychlorinated biphenyls and dioxins. Fish is the main source of methylmercury for the general population, thus there is a potential harm from mercury present in fish, especially for women of childbearing age and children.

Chapter II - Eating fish, especially cold-water ocean fish, has traditionally been associated with better health, due to generally high amounts of omega-3 fatty acids, vitamin D, and high-quality amino acids. However, due to reduced availability of such fish and an ever increasing world population, many people cannot obtain such fish. Thus, this chapter explores the possibility that the health benefits of eating fish could be obtained from largely vegetable sources. There are good vegetable sources of omega-3 fatty acids, including rapeseed (canola) oil, and flaxseed oil. Vitamin D3 (cholecalciferol) supplements are a reliable source of vitamin D, and both solar and artificial UVB are also good sources in moderation. Vegetable products can provide adequate amounts of protein and essential amino acids, with combining several types of vegetable products indicated for obtaining the essential amino acids in the ratio in which the body can use them.

Chapter III - To understand consumption patterns for self-caught fish it is essential to examine the knowledge base used by people in making decisions, including both the risks and benefits. Public perceptions of warnings about contaminants in fish have been frequently examined from a number of different locations, but seldom have both the risks and benefits been examined together, especially from different nearby locations. In this chapter I examine perceptions of the risks, warnings about, and benefits of fish consumption in three coastal estuaries. Fishermen and other recreationists (N = 329) were interviewed in Barnegat Bay New Jersey, the New York/New Jersey Harbor, and along the northern shore of Long Island in 2007 and 2008. Overall, a higher percentage of people from all three regions had heard about the benefits of fish consumption compared to the risks (or warnings about fish consumption). While there were no locational differences in having heard about the risks from fish consumption or in belief in any warnings, there were significant differences in knowledge about warnings, differences in contaminant loads among fish, and health benefits from eating fish. People from Long Island were less aware of warnings about fish consumption, and fewer knew that contaminants loads differed among fish, although the fewest people from the harbor were aware of the benefits of fish consumption. There were several other major differences: 1) awareness of mercury was higher in people interviewed from Barnegat Bay compared to elsewhere, 2) Word of mouth was they method over 55 % of the people from Barnegat Bay heard about warnings compared to 13 % from the harbor (and none on Long Island), 3) More people from Long Island had heard about warnings for bluefish and striped bass compared to those from elsewhere, 4) A higher percentage of people from the harbor knew about contaminants in bottom-dwelling fish compared to elsewhere, and 5) Fewer people from the harbor knew that fish were generally healthy and had high Omega-3 oils. These data illustrate that even people from the same region do not have the same knowledge base and that benefits of fish consumption are more widely known generally than the risks, although specific information about benefits is generally low. Public policy makers, risk managers, and health professionals need to be aware of the knowledge base of local residents, and a targeted information campaign is necessary to more broadly inform the public about the specific warnings (relative to both contaminants and fish species) and health benefits of fish consumption so that they can make informed decisions about consumption.

Chapter IV - Fish is an important part of a healthy diet. There are many nutritional benefits from including fish in the diet. Fish are low in saturated fats and contain high quality protein and omega-3 fatty acids (a class of fatty-acids that are essential for cell development and have other beneficial health effects). Fish also supply a number of vitamins and minerals including calcium, iron, zinc, vitamin A, niacin, vitamin B6 and vitamin D. Eating fish can contribute to a healthy heart and a child's proper growth and intellectual development. Studies have shown that eating fish regularly (such as twice per week) can reduce the chance of death from a heart attack. However, concerns have long existed about whether fish are safe to eat due to the trace amounts of chemicals that may be present in fish tissue. Low levels of natural and man-made chemicals are in the water in which fish live. Three of the chemicals that can accumulate in fish tissue are mercury, polychlorinated biphenyls (PCBs), and chlordane among others. Human health studies have indicated that reproductive function may be disrupted by exposure to these chemicals; neurobehavioral and developmental deficits occur in newborns and continue through school-aged children who had in utero been exposed. Other systemic effects (e.g., self-reported liver disease and diabetes, and effects on the thyroid and immune systems) are associated with elevated serum levels; and increased cancer risks are also associated with exposures. This article examines the economic importance of fish consumption in relation to good health and looks at the potentials of producing safe fish through aquaculture.

Chapter V - The patients with end-stage renal disease (ESRD) include significant percentage of malnourished patients together with other risk factors: dyslipoproteinaemia, insulin resistance, increased oxidative stress, inflammation, and they all together impair endothelial function.

Nutrition, glicoregulation disorders and insulin resistance as etiologic factors participate a grate deal in premature atherosclerosis. Abnormal fatty acid patterns are reported in patients with such diseases as diabetes mellitus, inflammatory bowel disease, rheumatoid arthritis and atherosclerosis. Basic mechanism of these disorders is connected with changes in cells functions on the membrane level. These basic discoveries help us in understanding the damage process of endothelial and appearance of its dysfunction by inflammation process and by lipid peroxidation process.

The authors examine association between atherosclerotic risk factors and nutritional status in hemodialysis patients. Their research is a part of the study in which haemodialysis patients will be followed. In the authors' previous study they examined HD patients which were administered supplementation n-3 PUFAs for two months. Dietary regime with fish oil could be used in dialysis patients to slow down atherosclerosis process and improve nutritional parameters.

Forty three hemodialysis patients were examined (20 males, 23 females, 55±12 years). Nutritional and inflammatory markers-including serum concentrations of C-reactive protein, tumor necrosis factor alpha, and interleukin 6- were measured. Mortality was followed prospectively for up to 12 mo.

Significant negative correlation between serum level of CRP and macroangiopathy was found. There was significant positive correlation between arachidonic acid (AA) and high sensitivity C-reactive protein (hs-CRP). Significant correlation was found between body mass index (BMI) and n-3 fatty acids.

Increasing inflammation and endothelial dysfunction predict the development of vascular disease. We report the relationship between inflammatory markers and nutritional parameters indices of atherosclerosis and other cardiovascular risk factors in patients on hemodialysis.

Chapter VI - *Background:* Epidemiologic data suggest that omega-3 fatty acids derived from fish oil reduce cardiovascular disease.

Objective: To assess whether dietary supplements of eicosapentaenoic acid (EPA) and docosahexaenoic acid (DHA) alter cardiovascular deaths and cardiovascular events

Data Sources: MEDLINE, Embase, the Cochrane Database of Systematic Reviews and citation review of relevant primary and review articles.

Study Selection: Prospective, randomized, placebo-controlled clinical trials that evaluated clinical cardiovascular endpoints (cardiovascular deaths and all cardiovascular events) in patients randomized to EPA/DHA or placebo. The authors only included studies that used dietary supplements of DHA/EPA which were administered for at least one year.

Data Extraction: Data were abstracted on study design, study size, type and dose of omega-3 supplement, cardiovascular events and duration of followup. Studies were grouped according to the risk of cardiovascular events (high risk and moderate/low risk). Meta-analytic techniques were used to analyze the data.

Data Synthesis: The authors identified nine studies that included 31 996 patients. Dietary supplementation with omega-3 fatty acids significantly reduced the risk of cardiovascular deaths (OR 0.8; 0.71-0.95, p=0.007). The mortality benefit was largely due to the studies which enrolled high risk patients. Similarly, dietary supplementation with omega-3 fatty acids significantly reduced the risk of all cardiovascular events (OR 0.82; 95% CI 0.74 to 0.90, p< 0.0001). This benefit was present in both high and low risk groups. Meta-regression failed to demonstrate a relationship between the daily dose of omega-3 fatty acid and clinical outcome.

Conclusions: Dietary supplementation with omega-3 fatty acids should be considered in patients at risk for cardiovascular disease.

Chapter VII - *Introduction:* In the past few decades, there has been an enormous interest in quantifying the relationship between fish consumption and the risk of various malignancies. Previous research suggested that consumption of even relatively small amounts of fish is associated with a reduced risk of several cancers, especially of oral cavity and pharynx, esophagus, stomach, colon, rectum, and pancreas. Despite the growing body of literature in this area, studies have reported mixed results. The aim of this chapter is to provide a comprehensive summary of evidence that exists in the area of fish consumption and cancer by concentrating on four major cancers described in relation to fish consumption, including colorectal, breast, prostate, and lung cancers.

Methods: Articles published from 1970 to 2008 were identified using the Pubmed search engine, and selected if they included results on the effect of dietary fish consumption on four major cancers, if they were cohort or case control studies, they were published in English or had expanded abstract available in English language. Case studies and studies with less than 10 participants were excluded. Two reviewers independently reviewed the articles using a structured abstraction form.

Results: Over 5,000 titles dealing with fish consumption and cancer have been identified through the initial search. Flagging the article titles that specifically dealt with the cancers of interest narrowed the search to approximately 50 studies in the area of biomarkers and 300 in the area of fish consumption and cancer. Seven studies in the area of biomarkers and fish consumption were evaluated in detail. These studies demonstrated that fish consumption plays an important role in inflammatory processes associated with cancer development.

Conclusions: Large number of cohort and case control studies from around the world provide interesting data about the efficacy of fish consumption for cancer prevention, however these data are still not conclusive. In particular, the effect of omega-3 fatty acid consumption on cancer incidence needs further investigation. There needs to be more well-designed longitudinal studies in this area to quantify the relationship between fish consumption and cancer.

Chapter VIII - Fish consumption confers nutritional benefits and potential contamination risks. The issue is of particular concern to reproductive-aged women and their infants. Although federal and local fish consumption advisories are controversial and appear to cause confusion, their behavioral and biological impacts are not well understood. The authors review available literature regarding correlates of fish consumption, the extent of advisory awareness among Americans, and the behavioral impacts of advisory awareness. They estimate advisory compliance and its sociodemographic and behavioral correlates among pregnant participants of a western Washington cohort study. Finally, they identify gaps in current knowledge and suggest areas for future research. Many studies show that maternal fish consumption directly relates to maternal and fetal mercury exposures. Far fewer describe women's knowledge and behaviors regarding the risks of fetal mercury exposures. Among 2,383 pregnant women in western Washington, lack of periconceptional fish consumption was associated with maternal age <20 years and both leanness and overweight before pregnancy. Consumption above federal advisory guidelines was associated with age ≥ 35 years, smoking before pregnancy, post-high school education, and nulliparity. Until we better understand the behavioral impacts of federal and local advisories, we will be unable to improve the ways in which we communicate the benefits and risks of fish consumption to reproductive-aged women.

Chapter IX - Shellfish is one of the main commercial productions in aquaculture industry of USA, Europe, Japan and China. Since shellfish feed by filtration, often they accumulate (bio-) toxins, bacteria, and parasites. As a result thousands people suffered Paralytic-, Amnesic-, or Diarrheic Shellfish Poisoning (PSP, ASP, DSP), ciguatera and other diseases during the twentieth century (and as consequence many people died). Relevant efforts have been made in prevention programs, treatments and legislation to reduce the incidence of such diseases (usually obtaining satisfactory results).

However, new emerging health problems are being linked to shellfish consumption. For example, mussels (Mytilus galloprovincialis) and cockles (Cerastoderma edule) store infectious Cryptosporidium parvum oocyst, which produce diarrhea in humans (being a relevant cause of died in AIDS infected people). On the other hand, European legislation allows less than 0,16μg DSP toxin (okadaic acid equivalent) per gram of meat (Decision 2002/225/CE), which is enough to prevent Diarrheic Shellfish Poisoning. In spite of this, DSP toxins are also potent tumor promoters. An epidemiological study shows a statistically significant correlation ($p<0.001$) between consumption of molluscs and the incidence of colorectal cancer (an increase of 7 times in shellfish consumption produced duplication in the risk ratio of colorectal cancer. Many other risks have not been taken in account. For instance, shellfish from estuaries can accumulate microcystins (potent hepatotoxins) produced by freshwater cyanobacteria transported to the estuary by the river. Microcystins have been implicated in the high rates of hepatic cancer in China population.

As a conclusion, a lot of work is still necessary to assure a safety in shellfish consumption.

Chapter X - Until recently, all patients receiving parenteral nutrition (PN) received all or a portion of their fat calories from lipid emulsions derived from soybean oils. It was felt that using another fat source alone, such as fish oil, would predispose patients to developing essential fatty acid deficiency (EFAD), bleeding, or other complications. This review describes recent developments in the use of fish oil based intravenous fat emulsions as monotherapy in the PN dependent patient as it relates to the reversal of PN associated liver disease and provides an overview of why complications such as bleeding or EFAD are rare. Other benefits of using fish oil as the preferred parenteral lipid source, such as prevention of retinopathy of prematurity and hypertriglyceridemia, will also be discussed.

Chapter XI - Cereals, of the taxonomic family Gramineae, are responsible for a high rate of allergic and immunologic reactions. Cereals include wheat, barley, rye, oats, rice and maize, and different consumption patterns of various cereals around the world may justify the different rates of sensitization reported for each cereal. They are the major components of the human diet in many countries around the world, and sensitization to any of the cereals consequently represents a major public health challenge. In many countries, corn (Zea mays L) is the basic ingredient of frequently eaten foods like porridge, polenta, corn flakes, popcorn, tacos, chips, canned de-embryonated kernels, and other food products. Corn is also component of products used in the elaboration of bread, cakes and beer. In addition, corn may be used in the elaboration of starch, glue, cosmetics, stickers, stamp adhesives, exipients (for example, in the formulations of aspirin and vitamins), chewing gum, talcum powder, lozenges, ointments, toothpastes, laxatives and bath powder [3]

The protein composition of maize seeds has been investigated in depth. However, few studies have addressed the identification of maize allergens. A major lipid transfer protein (LTP) allergen of maize appears to be a protein belonging to group 1 of grass pollen allergens that are responsible for seasonal respiratory symptoms; the maize allergen appears to be responsible for provoking food-induced allergen reactions. Similar to flour, particles of corn can also be inhaled by people handling it while making cakes, pastry cooks, by housewives and by farmers. Corn flour contains about 7% to 13% protein that can be fractionated into several solubility classes; the salt-extractable fraction (albumins and globulins) averages less than 4% of the total endosperm protein and mainly comprises of proteins with metabolic functions whereas 60-70% of proteins are called zeins.

Corn is often cited as an allergenic food and positive specific IgE to this cereal have been reported. In Mexico, despite the high consumption pattern of corn, there is only limited information on food-induced allergic reactions to maize in adults. Symptoms may include urticaria, diarrhea, wheezing, and in some cases anaphylaxis. Therefore, the authors performed the present study in order to evaluate specific IgE levels as a diagnostic test of corn allergy in children.

In: Fish Consumption and Health
Editors: George P. Gagne and Richard H. Medrano
ISBN 978-1-60741-151-2

Chapter I

Fish Consumption and Health

Christos Pitsavos[1], Christina-Maria Kastorini[2] and Christodoulos Stefanadis[1]
[1]First Cardiology Clinic, School of Medicine, University of Athens, Athens, Greece
[2]Dept. of Nutrition and Dietetics, Harokopio University, Athens, Greece

Abstract

Fish is one of the main components of a healthy diet and an excellent source of numerous beneficial compounds, such as proteins, vitamins, minerals and especially n-3 fatty acids, with several health benefits and particularly as far as cardiovascular disease is concerned. There is good evidence that modest fish consumption of approximately two servings of fish per week lowers the primary risk of coronary artery disease. Fish oil is rich in n-3 fatty acids which have anti-inflammatory, antithrombotic, antiarrhythmic, hypolipidemic, hypotensive and vasodilatory properties. Fish consumption also has beneficial effects in type 2 diabetes, stroke, renal disease, types of cancer and especially breast, endometrial, colorectal, renal and prostate, rheumatoid arthritis, asthma and chronic obstructive pulmonary disease. Furthermore adequate fish consumption is necessary in order to ensure proper brain development and cognitive performance. Deficiencies in n-3 long chain fatty acids or imbalance in the n-6 to n-3 fatty acid ratio have been implicated in a variety of central nervous system disorders, such as Alzheimer's disease and psychiatric illnesses, including depression and mood disorders. On the other hand fish consumption in large amounts has been related to health risks, mainly due to its content in toxic heavy metals, polychlorinated biphenyls and dioxins. Fish is the main source of methylmercury for the general population, thus there is a potential harm from mercury present in fish, especially for women of childbearing age and children.

Introduction

Fish is an important part of a well balanced healthy diet, as it provides protein of high biological value and low fat content, as well as the essential unsaturated fatty acids, certain vitamins and minerals. Fish contains the n-3 polyunsaturated fatty acids and especially eicosapentaenoic acid, EPA and docosahexaenoic acid, DHA that appear to play several beneficial roles for human health. In addition fish is the best dietary source for vitamin D and a good source of other vitamins like A, B3, B6, B12, E, minerals such as calcium, phosphorus, iron and trace elements like selenium, iodine and zinc [1-4].

The essential long chain n-3 polyunsaturated fatty acids, eicosapentaenoic acid (EPA) and docosahexaenoic acid (DHA), are found mainly in fatty fish. Nevertheless it should be kept in mind that the content of marine n-3 polyunsaturated fatty acids varies between fish species and depends on the location and season of capture even in the same species. Fatty fish are a rich source of n-3 fatty acids, while they have a lower content of saturated fat than many meat servings. Lean fish provide low amounts of n-3 fatty acids, but on the other hand contain little saturated fat and cholesterol [5, 6]. The n-3 fatty acid recommendation to achieve nutritional adequacy, defined as the amount necessary to prevent deficiency symptoms, is 0.6–1.2% of energy for alpha-linolenic acid; up to 10% of this can be provided by EPA or DHA [7]. Nonetheless, other components present in fish might also contribute to the beneficial effect of fish consumption [6].

Extensive scientific research has documented numerous health, social, cultural and economic benefits of eating fish. More specifically n-3 fatty acids, present in fish, play a vital role for the development and function of the nervous system and most importantly for the neonatal and infant brain development, as well as the photoreception and the reproductive system [1]. It appears that regular consumption of fish is suggested to reduce the risk of all-cause mortality, cardiovascular disease, coronary heart disease and myocardial infarction, decrease mild hypertension, prevent certain cardiac arrhythmias, sudden death and stroke and lower the incidence of diabetes, asthma and kidney disease [1, 8, 9]. Additionally, fish consumption has been linked to the prevention of several types of cancer [10, 11]. Finally fish consumption seems to offer protection against dementia, Alzheimer's disease, neuropsychiatric disorders and inflammatory disorders like chronic respiratory disease, rheumatoid arthritis and ulcerative colitis [2, 13].

On the other hand however, the overconsumption of certain types of fish have been linked to neurologic deficits, cancer, autoimmune and endocrine disorders, and even some heart disorders. The risks stem mainly from mercury, which accumulates over the lifetime of larger, longer-living fish, polychlorinated biphenyls, which are found in fish living in polluted waters and in some farmed fish and several other halogenated persistent organic pollutants [8, 13].

1. All Cause Mortality

Fish consumption is suggested to be beneficial as far as all cause mortality is concerned. In an ecological study fish consumption was associated with a reduced risk from all-cause, ischemic heart disease and stroke mortality across 36 countries [9]. Furthermore according to the NHANES I Epidemiologic Follow-up study, white men aged 25–74 years who consumed fish one time per week had an age-adjusted risk of non-cardiovascular death only about three quarters that of men never consuming fish. However no additional reduction in risk was seen for consumption more than 1 time per week and no consistent association of fish consumption and coronary heart disease incidence or mortality was seen [14]. A prospective cohort study from Finland that followed 415 patients with coronary artery disease for 5 years, reported a significant decrease in all-cause mortality in patients who consumed more than 57 gr fish per day compared to those who did not [15]. Finally according to the Diet And Reinfarction Trial (DART), a randomized controlled trial of n-3 fatty acids in the secondary prevention of coronary heart disease, a 29% reduction in all-cause mortality over a 2-year period in male myocardial infarction survivors advised to increase their intake of oily fish from 200 to 400 gr of fatty fish per week, has been reported. The greatest benefit was seen in fatal myocardial infarctions, and this observation led to the hypothesis that n-3 fatty acids might protect the myocardium against the adverse sequela of acute ischemic stress [16-18].

2. Cardiovascular Disease

The concept that fish intake may reduce the risk of coronary heart disease originated almost 30 years ago from reports on the small population of non-acculturated Eskimos in arctic Greenland with high consumption of marine animals [19, 20] and on inhabitants of the Japanese island of Okinawa who also consume high amounts of fish. In addition a large number of epidemiologic studies have found beneficial effects of fish intake on all cause mortality, coronary heart disease events, coronary heart disease death and sudden cardiac death, as well as ischaemic stroke and congestive heart failure [5, 15, 21, 22].

2.1. Coronary Heart Disease Mortality

There is a significant number of studies indicating that eating fish once per week or more frequently, is beneficial against coronary heart disease mortality. In a meta-analysis of prospective cohort studies, a consistently inverse association between fish consumption and coronary heart disease mortality rates has been found. More specifically, eating fish once per week might significantly reduce death from coronary heart disease by 15%. A dose-response relation has been observed, as an increment of 20 gr per day of fish intake could possibly lower coronary heart disease mortality rates by 7%. This inverse association was more apparent among studies with a follow-up period of longer than 12 years [23]. Furthermore a recent prospective cohort study with a 16-year follow-up, conducted with women in the Nurses' Health Study, reported an inverse association between fish intake and n-3 fatty acids and coronary heart disease death. Compared to women who ate fish less than once per month, the risk of coronary heart disease death was 21%, 29%, 31%, and 34% lower for fish consumption 1 to 3 times per month, once per week, 2 to 4 times per week, and >5 times per week, respectively [24]. In a prospective study of men of three European cohorts from the

Seven Countries Study, no association has been shown between total fish consumption or lean fish consumption and coronary heart disease mortality. However fatty fish consumption compared to non-fatty-fish consumption was associated with a 34% reduction in 20-year coronary heart disease mortality, suggesting that especially fatty fish is protective against death from coronary heart disease [25]. Across different studies, compared to little or no intake, modest fish consumption lowers relative risk of coronary heart disease mortality by 25% or more. Higher intakes do not substantially further lower the risk, suggesting a threshold of effect [12]. In another study, fish consumption has been found to be associated with a decrease in the relative risk of coronary heart disease mortality by 17% compared to consumption of fish less than one serving per month, with each additional serving per week associated with a further reduction in this risk of 3.9% [26]. Additionally fish consumption versus little to no fish consumption has been associated with an approximately 20% reduction in the risk of fatal coronary heart disease and a 10% reduction in total coronary heart disease, possibly due to the beneficial effects of n-3 fatty acids in reducing sudden cardiac death [27]. Consumption of 35 gr of fish per day has been found to be inversely associated with death from coronary heart disease, especially non-sudden death from myocardial infarction, as men who consumed 35 gr or more of fish per day had a 42% lower rate of death from myocardial infarction than the non-consumers. The age-adjusted rates of death from myocardial infarction, coronary heart disease, cardiovascular diseases, and all causes were lowest for the men with the highest consumption of fish, with a trend toward lower mortality rates with higher levels of consumption. Differences in death from coronary heart disease, cardiovascular diseases and all causes among the four strata of fish consumption were attributable to differences in death from myocardial infarction, especially non-sudden death, as fish consumption was not significantly associated with the risk of death from other coronary causes [28]. It should be also mentioned than middle-aged or older men in Shanghai, China, who consumed at least one serving of fish and shellfish per week had a 44% reduced risk of fatal myocardial infarction compared with less-frequent consumers, demonstrating an inverse association between seafood consumption and total mortality. Men who consumed more than 200 gr of fish or shellfish per week had a 59% reduction of fatal acute myocardial infarction compared with men consuming less than 50 gr per week and approximately a 20% reduction in total mortality [29]. Very recently, fatty acid compositional data on plasma phospholipids from the Cardiovascular Health Study, drawn 2 years before a cardiac event, exhibited a 70% lower odds ratio for fatal ischemic heart disease in those with higher blood levels of EPA plus DHA [30]. However according to another study, although the associations between EPA and DHA and the risk of death were confirmed by the reduced risk observed in the subjects who ate fish or who had high proportions of EPA and DHA in serum phospholipids, the associations of n-3 fatty acids with combined fatal and nonfatal cardiovascular disease events were not significant [31]. Finally consumption of broiled or baked fish 3 or more times per week has been associated with 49% lower risk of total ischemic heart disease death and 58% of arrhythmic ischemic heart disease death, but not nonfatal myocardial infarction, compared with consumption less than once per month. [32].

2.2. Acute Coronary Syndromes

Observational studies have found inverse relations between fish consumption and risk of developing coronary heart disease, suggesting that populations consuming more n-3 fatty acids from fish have a lower incidence of coronary artery disease [33]. Compared to a modest fish intake of once a week, higher intake was associated with substantially reduced risk of coronary heart disease, primarily non-fatal cardiac events, among middle-aged persons. High consumption of fish, 8 times per week has been associated with reduced risk of coronary heart disease, more specifically, myocardial infarction and non-fatal coronary heart disease, compared to a modest fish consumption, of once a week, suggesting that a high fish intake may have a further beneficial effect for the prevention of coronary heart disease among middle-aged persons [34]. Moreover the protective effect of small amount of fish consumption in relation to coronary heart disease observed in middle-aged people seems to be also present in the elderly [35]. In the USA, high consumption of fish was associated with lower risk of coronary heart disease in women, and with lower coronary heart disease incidence and total mortality in diabetic women. In the NHANES I study, moderate fish consumption once a week was also found to have a protective effect, but no further reduction in risk was observed in higher fish consumption [36]. Additionally various studies have indicated that long-term fish consumption, appears to be associated with lower primary and secondary heart attack rates from cardiovascular disease [30]. Moderate fish consumption, less than 150 gr per week, was associated with 38% lower odds of developing acute coronary syndromes as compared to no consumption and a J-shape association between fish intake and coronary risk has been suggested. The beneficial effect from low fish consumption was also observed even among current smokers and diabetics [36-38]. Furthermore, small quantities of fish consumption have been associated with risk reductions in non-fatal myocardial infarction risk by 27%, but additional fish consumption conferred no incremental benefits. As a result, the impact of fish consumption on nonfatal myocardial risk is very uncertain [26]. There is a strong inverse association seen between the risk of a first myocardial infarction and the biomarkers of fish intake, Ery-Hg, which reflects intake of all fish and P-polyunsaturated fatty acids, an index of intake of fatty fish, independent of traditional risk factors [39]. N-3 PUFA and fish consumption have been inversely associated with non-fatal acute myocardial infarction in Italy. Patients eating fish less than once per week had an OR of 0.80, compared to those eating fish less than twice per week, and those eating fish twice or more per week had an OR of 0.68, with a significant trend in risk [40]. Chronic intake of fish was associated with a reduction in infarct size as estimated by the frequency of Q-wave infarcts after myocardial infarction [41]. In a prospective cohort study of diabetic women, higher consumption of fish and n-3 fatty acids was associated with a lower incidence of both coronary heart disease and total mortality [42]. Finally compared to no dietary intake of eicosapentaenoic acid and docosahexaenoic acid, intake of the equivalent of one fatty fish meal per week, was associated with a 50% reduction in the risk of primary cardiac arrest [43].

2.3. Cardiac Arrhythmias - Sudden Cardiac Death

Fish intake seems to provide antiarrhythmic protection at a population level, as long-term consumption of fish is strongly and inversely associated with QTc duration on resting electrocardiographic recording in free-eating people without any evidence of cardiovascular disease. In particular compared to fish non-consumers, those who consumed more than 300 gr of fish per week had a mean 13.6% lower QTc and a 29.2% lower likelihood of having QTc intervals >0.45 s [44]. Furthermore greater habitual consumption of tuna or other broiled or baked fish was associated, in a graded fashion, with significantly lower heart rate. Compared to intake less than 1 month, consumption 3 times per week was associated with 3 bpm lower heart rate. Evidence also suggests that fish consumption improves left ventricular diastolic function, possibly due to the beneficial effects of n-3 PUFAs [22].

Fish consumption has been shown to be related to reduced sudden cardiac death. Patients with coronary artery disease who eat fish, appear to have a lower subsequent rate of coronary artery disease and lower total mortality, especially from sudden death [33]. In a population-based, nested, case-control study, a strong negative relationship was reported between fish intake and risk for sudden death. In particular, two fatty fish meals per week were associated with a 50% reduced risk of primary cardiac arrest [43]. In addition a very strong inverse relation between the summed blood levels of the long-chain n-3 PUFA and the risk of sudden cardiac death has been reported in men without prior evidence of cardiovascular disease. Higher blood DHA levels were found to be associated with a higher heart rate variability which may contribute to the lower risk of cardiac arrhythmia and sudden death with higher intakes of fish derived fatty acids [30]. According to the US Physicians' Health Study, men who consumed fish at least once weekly had a relative risk of sudden death of 0.48 versus men who consumed fish less than once per month [45].

2.4. Stroke

Strokes fall into two main categories: ischemic and hemorrhagic. Ischemic strokes are caused by a blockage of arteries leading to the brain and they result from atherosclerosis of the carotid and cerebral arteries plus thrombosis. Hemorrhagic strokes result from a weakened blood vessel that ruptures and bleeds into the surrounding brain tissue [46]. The results regarding fish consumption and stroke are conflicting and while some studies show a beneficial effect of fish consumption [46-53], others indicate increased risk of stroke for subjects consuming the highest amounts of fish [54, 55].

The evidence to date is primarily for total stroke risk, and the associations could differ if the data were analyzed for type-specific stroke incidence. Evidence exists for an inverse relationship between small intakes of fish and ischemic stroke and for a possible increased risk for hemorrhagic stroke with "Eskimo" intakes of n-3 fatty acids. Thus, linking fish consumption with total stroke risk is likely to underestimate the strength of the real associations between fish consumption and type-specific stroke risk [16].

Most studies show a beneficial effect of fish consumption and stroke. Even consumption of at least one portion of fish per week may be associated with a reduced stroke incidence

[49] and especially ischemic stroke [46]. In the Nurses Health Study, the relative risk of total stroke was lower, although not significantly, among women who regularly ate fish than among those who did not. A significant 51% decrease in the risk of thrombotic stroke was observed among women who ate fish at least two times per week, compared to women who ate fish less than once per month [47]. In addition in a large prospective study among male health care professionals, a 40% lower risk of ischemic stroke was observed in men who consumed fish once per month or more compared to those who ate fish less often. However, a higher frequency of fish intake was not associated with further risk reduction, indicating that the beneficial effect of fish intake on risk of ischemic stroke might be maximal at a relatively low level of consumption [50]. According to a recent meta-analysis of cohort studies, an inverse association has been found between fish consumption and risk of stroke and particularly ischemic stroke, indicating a significant reduced incidence of ischemic stroke by consuming fish as seldom as 1 to 3 times per month. However there was not a strong dose-response relationship observed. Compared to those who never consumed fish or ate fish less than once per month, individuals with higher fish intake had lower risk of total stroke. In fact the reduction in risk of total stroke was statistically significant for fish intake once per week [51]. In addition consumption of fish 2 to 4 times per week was associated with a reduced risk of total stroke and thrombotic infarction, primarily among women who did not take aspirin regularly [52]. Moreover according to the Physicians' Health Study, men consuming more than 5 servings of fish per week had a non-significantly lower risk of fatal and nonfatal stroke compared with men consuming less than one serving per week [56]. Finally, fish consumption was inversely associated with total stroke and ischemic stroke, with 27% lower risk of ischemic stroke with an intake of 1 to 4 times per week and 30% lower risk with intake of 5 or more times per week compared to an intake of less than once per month [53].

There are also studies that do not show significant associations between fish consumption and incidence of stroke. In a prospective study of a British population, with average follow-up of 8.5 years, no consistent relationship between fish consumption and stroke was observed. However, there appeared to be an inverse association between oily fish consumption and the relative risk of stroke in women [57]. Additionally ten year stroke-related mortality was significantly lower in persons consuming fish at least once per week compared to less frequent consumption. However the long-term protective effects on stroke mortality associated with eating fish were attenuated after incorporating the adjustment for retinal vessel diameters [58].

On the other hand there are some studies that show an increased risk of stroke for those at the highest levels of fish consumption. According to the Chicago Western Electric Study, men consuming the highest level of fish, 35 gr per day, had a higher age-adjusted death rate from stroke, than men in the three other categories of fish consumption [54]. In addition according to a more recent study those in the highest quintile of consumption, 46 gr of fish per day, had 95% greater possibility of stroke when compared to those in the lowest quintile of fish consumption, 11 gr per day. The risk of cerebral infarction also increased with the consumption of fish. Therefore these findings suggest that high fish consumption is associated with increased risk of stroke and cerebral infarction [55]. Finally, according to another study only fried fish or fish sandwich consumption was positively associated with total stroke and ischemic stroke, with a 44% higher risk of ischemic stroke with consumption of more than once per week compared with consumption of less than once per month [53]. However, even if fish intake may be a risk determinant for ischemic stroke, reductions of fish consumption should not be recommended [59].

At this point it should be mentioned that gender and race might affect the relationship between stroke risk and fish intake. The effect of fish intake on stroke risk seems to be different in men and women. Fish intake seems to be associated with increased risk of ischaemic stroke only in men, but the effect appears to be small compared with the protective effect on myocardial infarction [59]. In addition according to another study, white women who consumed fish more than once a week, had a risk of stroke incidence only about half of that of women who never consumed fish. However fish consumption more than once a week compared with never, was not associated with age-adjusted stroke risk in white men. In black women and men, however, any fish consumption compared with never was significantly associated with reduced adjusted stroke risk [48]. Inconsistencies in the observed health effects of fish consumption in different populations may reflect different patterns and type of fish consumed and preparation methods [57].

The observation of high incidence of hemorrhage in Eskimos, a population consuming large amounts of fish, as well as biological evidence have raised concerns about possible adverse effects of high fish intake on risk of hemorrhagic stroke [51, 53], nevertheless this has not been verified by recent studies [47, 50, 52]. According to a recent meta-analysis, no significant associations were observed between fish intake and hemorrhagic stroke. This observation could be explained not only by the fact that hemorrhagic stroke represents only a modest proportion of all strokes [46], but also because the risk of hemorrhage caused by high fish consumption may be balanced by other benefits of fish intake [51].

2.5. Pathophysiological Mechanisms

The beneficial effects of fish are mainly attributed to the long-chain polyunsaturated n-3 fatty acids, which are reported to influence several cardiovascular risk factors and reduce various pathophysiologic abnormalities in heart and blood vessels. [12, 60, 61].

Cardiovascular Disease Risk Factors

Long-term fish intake has been associated with reduced levels of the most common cardiovascular disease risk factors not only in the general population, but also in a cohort of elderly people. Long-term increased fish consumption has been related to a significant reduction in systolic blood pressure, total serum cholesterol, triglyceride and fasting blood glucose levels, while no effect of fish intake was observed on diastolic blood pressure, HDL, and LDL cholesterol levels [16, 44, 62]. N-3 fatty acids are also suggested to reduce levels of VLDL cholesterol, improving the lipoprotein profile by decreasing the fraction of atherogenic small, dense LDL, not only in healthy adults but also among patients with hypertriglyceridemia and diabetics [38, 42, 63, 64].

N-3 fatty acids might prevent insulin resistance, indicating that high fish intake may reduce the risk of type 2 diabetes in populations with a high prevalence of obesity. In countries with low fish and seafood consumption, the prevalence of type 2 diabetes increased significantly with obesity, whereas in countries with a greater prevalence of obesity, there was evidence of significantly reduced prevalence of type 2 diabetes with high fish and seafood consumption [65, 66]. Higher fish intake has been also associated with lower risk of microalbuminuria in type 1 diabetic patients [42].

Incorporating a daily fish meal into a weight-loss regimen was more effective than either measure alone at improving 24-hour ambulatory blood pressure, awake ambulatory heart rates [67], glucose and insulin metabolism, dyslipidemia and at reducing leptin levels [68] in obese individuals being treated for hypertension. As far as blood pressure levels are concerned, a daily meal of fish rich in n-3 fatty acids into a reduced-fat, energy-restricted diet in obese treated hypertensive subjects resulted in additive effects on blood pressure reduction with effects being greatest on awake systolic and diastolic blood pressure. A fall of 13 mm Hg in systolic blood pressure and 9 mm Hg in diastolic blood pressure compared with pressures in control subjects was observed, which was over and above that already achieved by the antihypertensive drug therapy. The reduction in heart rate seen with dietary fish suggests a cardiac-autonomic component, as well as vascular effects, of increased consumption of n-3 fatty acids from fish [67]. In addition the effect of the combination of fish and weight loss on the glucose and insulin metabolism was of interest, since no significant effect of fish consumption in participants in the nonenergy-restricted groups was observed. Improved insulin sensitivity in the fish plus weight-loss group may be due to changes in fatty acid delivery to the liver as a result of a profound reduction in postprandial lipemia associated with n-3 fatty acids, as well as a reduction in abdominal fat cell mass [69]. Finally the fall in serum leptin was significantly associated with reductions in serum insulin, ambulatory blood pressure and heart rate. It is important to mention that higher leptin concentrations were prospectively implicated as an independent risk factor for stroke, coronary artery disease and myocardial infarction [68].

Platelet Aggregation

The essential n-3 fatty acids are suggested to decrease platelet aggregation, suppress the formation of fibrinogen, increase fibrinolytic activity, reduce the viscosity of the blood and leukocyte and platelet reactivity, resulting in a modest prolongation of bleeding times [16, 64].

Coronary Atherosclerosis

Fish consumption has been associated with a significantly reduced progression of coronary atherosclerosis. Compared to lower fish intakes, consumption of more than 2 servings of fish or more than 1 serving of tuna or dark fish per week, was associated with smaller increases in the percentage of stenosis, smaller decreases in minimum coronary artery diameter and fewer new lesions in diabetic, but not in non-diabetic postmenopausal women with coronary artery disease [70]. Fish consumption has been also suggested to be protective against early atherosclerosis in middle-aged men, probably through its beneficial effects on inflammation, through anti-atherosclerotic effect of n-3 polyunsaturated fatty acids. The average intima-media thickness in those who ate fish more often was thinner than that in those who ate fish less often, although none of the participants showed any overt atherosclerosis on carotid ultrasonic examination [66]. The above findings could be explained by the fact that eicosapentaenoic and docosahexaenoic acids are actually incorporated into the phospholipids and cholesterol esters of severe atherosclerotic lesions in humans, playing an actual role in the reduction of the progression of atherosclerosis. [33]. Furthermore the vascular protective effects of consuming fish could act by preventing pathological microvasculature change. Regular fish in the diet has been associated with reduced prevalence of two important focal microvascular signs, arterio-venous nicking, and retinopathy lesions in non-diabetic persons. Greater frequency of fish consumption was associated with wider retinal arteriolar diameters, narrower retinal venular diameters, and lower prevalence of arterio-venous nicking and non-diabetic retinopathy [58].

Inflammation

In addition regular fish consumption attenuates inflammatory processes by antagonizing pro-inflammatory cytokines that increase risk of cardiovascular disease [58]. It is suggested that the principle mechanism of cardioprotective and neuroprotective action of n-3 fatty acids may be due to the suppression of TNF-a, interleukin-1 (IL-1) and IL-2 synthesis and release, modulation of hypothalamic-pituitary-adrenal anti-inflammatory responses, and an increase in acetyl-choline release, the vagal neurotransmitter [64, 71]. Moreover EPA and DHA seem to alter the metabolism of adhesion molecules such as vascular cell adhesion molecule-1 (VCAM-1), E-selectin, and intercellular adhesion molecule-1 (ICAM-1). [16, 72].

More specifically fish consumption was independently associated with lower inflammatory markers levels, among healthy adults in a dose-response manner, as compared to non-fish consumers, those who consumed 300 gr of fish per week had on average 33% lower C-reactive protein, 33% lower interleukin-6, 21% lower tumor necrosis factor-alpha, 28% lower serum amyloid A levels, and 4% lower white blood cell counts [73].

2.6. Inverse Results

Nevertheless, in spite of the fact that most epidemiological studies have reported a protective effect associated with the consumption of even small amounts of fish, there have been negative reports, as well as inconsistent findings [21, 23, 74]. According to a review of prospective cohort studies, individuals at low risk of coronary heart disease do not gain additional protection against coronary heart disease by fish consumption. However, it was suggested that high-risk individuals appear to benefit from increasing their fish consumption up to the optimum of 40–60 gr/day [75]. Furthermore in a cohort study of middle-aged men and women, no protective effect could be attributed to fish consumption regarding all-cause mortality or incident coronary heart disease in the population as a whole, nevertheless it should be kept in mind that frequent consumption of fish benefits those at high risk for coronary heart disease. This observation could possibly be explained by the type of the preparation method, as in Denmark, fish is often consumed on open sandwiches with butter and mayonnaise or pan-fried served with a butter sauce [76]. Neither the Health Professionals' Follow-up Study [77], nor the US Physicians' Health Study [45], showed any significant associations regarding fish intake and risk of total cardiovascular mortality or any fatal coronary disease, including sudden and non-sudden cardiac death, fatal or nonfatal myocardial infarction, coronary artery bypass grafting, or angioplasty. In contrast, fish consumption was related to a reduced risk of total mortality [45]. Lack of association between fish intake and coronary heart disease incidence and mortality has been also reported by an analysis of the Seven Countries data and the EURAMIC (European Multicenter Case-Control Study on Antioxidants, Myocardial Infarction and Breast Cancer) Study [78, 79]. In the Seven Countries Study, although an inverse association between fish consumption and 25-year mortality from coronary heart disease across several populations was observed, when the confounding effects of saturated fatty acids, flavonoids, and smoking were considered, the association was not significant [78]. In the EURAMIC Study, a large international case-control study, no evidence of a protective effect of adipose tissue DHA, a measure of long-term fish consumption, on the risk of developing myocardial infarction was seen [79]. Moreover in Japanese men, fish consumption was not associated with the risk of non-fatal myocardial infarction, in both genders [80]. In addition a small increase in coronary heart disease mortality in a study of Finnish men has been observed for those who consumed >40 gr/day of fish, mostly lean, but a small decrease for those who consumed fatty fish [25]. Finally in a recent study of 3114 men below 70 years of age with stable angina pectoris by the investigators of the DART, patients advised to eat fatty fish twice a week for 3–9 years had an increased risk of coronary heart disease. In particular, total mortality increased by 15%, coronary deaths by 26% and sudden cardiac death by 54% in the fish advice group [5, 81].

The variability observed could be explained by differences in the methods of assessing fish intake, different distributions of reported fish intake, different overall dietary patterns between populations, consumption of different types of fish, fatty or lean or fish from various sources, potential contamination of fish by toxic heavy metals, the possibility of reverse causality, as people are more aware of the cardioprotective effects of fish consumption, differences in the methods used to validate and classify endpoints, with only a few studies reporting, for instance, on both sudden and nonsudden cardiac death, or differences in risk levels among different populations [21, 40, 74]. Comparing different types of fish, lower risk appears more strongly related to intake of oily fish, such as salmon, herring and sardines, rather than lean fish, like cod, catfish or halibut [12].

As far as the type of fish meal consumed and the preparation method are concerned, fried fish or fish sandwich consumption from fast food restaurants or supermarket frozen sections does not seem to lower the risk of total ischemic heart disease death, arrhythmic ischemic heart disease death, or nonfatal myocardial infarction. On the contrary there seems to be a trend toward higher risk, possibly because of the lower n-3 PUFA content of these products which are prepared mostly from lean fish. Furthermore frying increases the n-6 to n-3 ratio, while the presence of trans-fatty acids due to the use of partially hydrogenated oils and lipid oxidation products, because of the reused oils in multiple frying cycles, may increase cardiovascular risk. Finally other lifestyle factors could affect this relationship [12, 32]. Therefore the consumption of commercially prepared fried fish from restaurants and fast food establishments, as well as many frozen, convenience-type fried fish products should be avoided [16].

It has been also suggested that methylmercury present in fish may attenuate the benefits of fish consumption to at least some degree [26, 82, 83], as recent studies have produced conflicting results with regard to the effects of methylmercury on coronary heart disease risk [13, 84]. Therefore, the principal question may not be whether consumption of mercury containing fish increases cardiovascular risk, but whether consumption of such fish would decrease risk even further if mercury were not present [2, 12, 85]. High intake of mercury from non-fatty freshwater fish and the consequent accumulation of mercury in the body are associated with an excess risk of acute myocardial infarction, as well as death from coronary heart disease and cardiovascular disease in Eastern Finnish men [84]. Fish intake and hair mercury were associated with an increased risk of acute myocardial infarction and death from coronary heart disease, cardiovascular disease, and all-cause mortality. Men in the highest tertile of hair mercury levels had a twofold relative risk of acute myocardial infarction and a 2.9-fold relative risk of cardiovascular death compared with those with lower hair mercury levels [85]. However, another study found a positive relationship between fish consumption and toenail mercury levels, but no increase in risk of coronary heart disease with higher mercury levels even after adjusting for age, smoking, and other risk factors [86].

Mechanisms by which mercury is thought to increase the risk of coronary heart disease include the reduction of antioxidative capacity, promotion of free radical stress and lipid peroxidation in cell membranes and lipoproteins. Mercury as a transition metal not only participates in Fenton-type reactions, resulting in the formation of free radicals, but it may also inactivate antioxidative compounds such as glutathione, catalase, and superoxide dismutase. In addition, mercury forms an insoluble complex with selenium, thus binding selenium in an inactive form that cannot serve as a catalytic center of glutathione peroxidase. Other possible atherogenic effects of mercury may be the potentiation of ADP-induced platelet aggregation, promotion of blood coagulation, inhibition of endothelial cell formation and migration, as well as effects on apoptosis and inflammatory responses [13, 87].

3. Cancer

Epidemiologic investigations studying how intake of fish and fish fatty acids influence the occurrence of cancer have been inconclusive. More specifically according to the World Health Organization (WHO), there is inconclusive evidence that increased consumption of fish is effective in cancer prevention [88].

3.1. Colorectal Cancer

Consumption of even relatively small amounts of fish is a favorable indicator of the risk of several cancers, especially of the digestive tract and notably colon and rectal cancer, which are among the leading causes of cancer mortality in developed countries. In particular a consistent pattern of protection has been observed against the risk of digestive tract cancers with fish consumption for the highest compared with the lowest level of consumption: oral cavity and pharynx, esophagus, stomach, colon, rectum, and pancreas [11]. Fish consumption, and possibly n-3 fatty acid intake, has been suggested to inhibit colorectal carcinogenesis. Consumption of fish rich in n-3 fatty acids is suggested to reduce colorectal cancer risk through inhibition of the arachidonic acid cascade that plays a role in inflammation and has been linked to carcinogenesis. In a recent meta-analysis, a borderline significant 12% lower risk of colorectal cancer was found for those in the highest fish consumption category compared with those in the lowest consumption category. Extra consumption of fish once per week corresponds to a 4% lower risk of colorectal cancer. The association was slightly more pronounced in studies from the United States than in those from Europe [89]. Furthermore according to the Fukuoka Colorectal Cancer Study, a large case-control study in Japan, diets high in fish and n-3 fatty acids reduce the risk of colorectal cancer, particularly in the distal colon [90]. Finally intake of fish has been inversely associated with colorectal cancer risk, with the trend for an inverse association being statistically significant for cancers of the left side of the colon and the rectum, but not for cancers of the right side of the colon [91].

On the other hand however there are studies that do not support the previous findings, showing no associations between fish consumption and colorectal cancer [92-94]. In a study no protective effect of fish against colon cancer risk was found, as no association between fish consumption and risk of colon cancer was seen, whereas a somewhat higher, however not statistically significant, risk of developing colon cancer with higher intake of fish was observed [93]. According to a prospective study, consumption of fish did not increase risk of colon cancer, but evidence was found that people who consume red and processed meat frequently have higher risk of rectal cancer. Therefore a reduction of the risk of rectal cancer is possible by substituting red meat consumption by other types of meat, such as fish [94].

3.2. Breast Cancer

Studies examining the relations between fish consumption and breast cancer show conflicting results. There are studies that support a beneficial relationship [95-97], others that show no associations [3, 98, 99] and studies that reveal an increase in the risk of breast cancer with high fish consumption [100-102].

An inverse association has been observed between percent calories from fish and breast cancer rates, demonstrating that of the dietary components considered, percent calories from fish was the most strongly correlated factor with breast cancer rates after statistical adjustment for dietary fat intake [95]. Consistent with a protective effect, The Singapore Chinese Health Study, a prospective cohort study with 314 breast cancer cases found a significantly reduced risk of breast cancer for subjects consuming high amounts of fish and marine n-3 fatty acids [96]. Similarly, a significant inverse association between fish intake and breast cancer risk has been observed in Japanese women in a study of 2,385 breast cancer cases [97].

There are also studies showing no significant associations between fish consumption and breast cancer. High consumption of fish was only weakly associated with reduced breast cancer risk, with the association not reaching statistical significance. In addition when the type of fish was examined separately, the association was similar for fatty and lean fish [98]. Moreover no significant associations between intake of total fish and breast cancer risk were observed in the European Prospective Investigation into Cancer and Nutrition. When examining lean and fatty fish separately, a positive significant association was found but only in the highest quintile for fatty fish, however test for trend was not significant [3]. Finally no association was detected between the overall frequency of fish for dinner and breast cancer risk, but there was an inverse relation with the frequency of main meals containing fish in poached form [99].

Higher intakes of fish have been associated with higher incidence rates of breast cancer in a prospective study. More specifically the incidence of estrogen receptor positive tumors was increased for higher intakes of fish, whereas no increase was seen for estrogen receptor negative tumors. The association was independent of fat content and preparation method of the fish [100]. Additionally in a cohort study of 88,795 women, a slightly increased risk of breast cancer was found, associated with higher dietary marine n-3 fatty acids [102]. According to a case-control study a significant increase in breast cancer risk was observed for consumption of fried fish whereas a significant inverse association was found for fish that was not fried, indicating that the cooking method may be important. Heterocyclic amines and polycyclic aromatic hydrocarbons formed during preparation of the fish at high temperatures may be one of the reasons [101].

3.3. Endometrial Cancer

Total fish consumption was inversely associated with risk, although weakly, but consumption of fatty fish was inversely associated with endometrial cancer risk, indicating that consumption of fatty fish, but not other types of fish, may decrease the risk of endometrial cancer [11, 103].

3.4. Prostate Cancer

Most studies suggest that fish consumption could be associated with decreased risk of prostate cancer. Men eating fish more than three times per week had a lower risk of prostate cancer, with the strongest association observed for metastatic cancer. Each additional daily intake of 0.5 gr of marine fatty acid from food was associated with a 24% decreased risk of metastatic cancer. Marine fatty acids may account for part of the effect, but other factors in fish may also play a role [104]. In addition during 30 years of follow-up, men who ate no fish had a two-fold to three-fold higher frequency of prostate cancer than those who ate moderate or high amounts did. Fatty fish consumption could possibly lower the risk of prostate cancer, possibly through inhibition of arachidonic acid-derived eicosanoid biosynthesis [105]. According to a prospective cohort study, an inverse association has been observed between fish intake and the risk of prostate cancer during 7 years of follow-up [106]. Eating fatty fish once or more per week, compared to never, was associated with reduced risk of prostate cancer, indicating that frequent consumption of fatty fish appears to reduce the risk of prostate cancer. In contrast, intake of white fish or shellfish was significantly associated with an increased risk of prostate cancer. This association is being modified by a genetic variation in the COX-2 gene, as strong inverse associations with increasing intake of salmon-type fish among carriers of the variant allele, but no association among carriers of the more common allele. Interestingly, the differences in fish intake or the distribution of COX-2 polymorphisms could explain part of the international variation in prostate cancer incidence rates [107].

3.5. Lung Cancer

Whereas some studies have observed a protective effect [108, 109], particularly where fish consumption is more frequent than once per week [110], other studies have observed a positive association between consumption of fish and risk of lung cancer [111].

Fish consumption has been inversely and significantly associated with a reduced risk of lung cancer mortality, but this possible protective effect is clear cut only in men and in countries with high level of cigarette smoking [112]. Furthermore linearly decreasing relative risks for lung cancer with increased frequency of consumption of fish and shellfish were observed, but not with intake of dried or salted fish. Decreased relative risks were apparent with both broiling and boiling cooking methods, but reduction with raw and deep-fried fish consumption was not statistically significant [108, 113]

On the other hand, consumption of fish could exert an influence by increasing the risk of lung cancer, indicating a significant risk for total fish consumption more often than once per week. The highest risk was for blue fish, though there was also a non-significant association for white fish for consumption more often than once per week. For total fish consumption, a positive association very close to significance was observed and maintained for consumption of both white and blue fish. Fish, with substantially lower consumption frequencies, would appear to have certain influence on the appearance of lung cancer, regardless of the type of fish [111].

To explain the results obtained, the precise type of fish, fatty or lean and the cooking method must be taken into account, since species that register the highest risk, are usually broiled or grilled, thus increasing their polycyclic aromatic hydrocarbon content greater and so raising their carcinogenic potential [111].

3.6. Renal Cell Carcinoma

In a large population-based cohort study with data on long-term diet, women who consumed 1 or more servings of fatty fish per week, had a statistically significant 44% decreased risk of renal cell carcinoma compared to women who did not consume any fish. Women who reported consistent long-term consumption of fatty fish at baseline and 10 years later, had a statistically significant 74% lower risk. In contrast, consumption of lean fish or other seafood was not associated with the risk of renal cell carcinoma. Frequent consumption of fatty fish may lower the risk of renal cel carcinoma possibly due to increased intake of eicosapentaenoic acid and docosahexaneoic acid, as well as vitamin D [114].

4. Rheumatoid Arthritis

Dietary n-3 fatty acids may favorably influence the risk of developing rheumatoid arthritis. Greenland Eskimos have a low incidence of this disease, compared to the Danish population [115]. Moreover the Japanese, who have a high average consumption of long-chain n-3 fatty acids compared to Western norms, have approximately one-third the incidence of rheumatoid arthritis [1, 116, 117]. Consumption of broiled or baked fish, but not of other types of fish, has been associated with a decreased risk of rheumatoid arthritis. A case-control study has shown a reduced risk for developing rheumatoid arthritis in subjects consuming 2 or more fish meals per week with an adjusted odds ratio of 0.57 (95% CI 0.35 to 0.93) compared to subjects consuming less than 1 fish meal per week [118].

5. Pulmonary Function

Chronic dietary intake of fish has been associated with higher levels of pulmonary function and is consistent with the hypothesis of an effect of fish oil on arachidonic acid metabolism. In a sample of 2,526 adult subjects aged 30–70 years from the First National Health and Nutrition Survey, a monotonic relationship between dietary portions of fish per week and the natural logorithm of predicted forced expiratory volume in one second was found, as eating fish more than once a week was associated with a higher level of lung function, compared with consumption less than once a week [119]. In addition fish consumption is suggested to limit the damage to the lungs caused by cigarette smoking, as an inverse relation has been observed between fish consumption and chronic obstructive lung disease among current and past smokers [38]. The pathogenesis of chronic bronchitis, emphysema, and deterioration of lung function in smokers is likely to involve inflammatory processes. More specifically, cigarette smoking is associated with an accumulation of neutrophils in the lung, increased leukocyte production of leukotriene B_4, a potent inflammatory metabolite of arachidonic acid, and enhanced release of reactive oxygen metabolites by alveolar macrophages [120]. Therefore fish consumption seems to be beneficial due to the inhibition of inflammatory mechanisms by the n-3 fatty acids in the lung [38].

5.1. Asthma

Asthma is now considered as a chronic inflammatory airway disease and therefore it is possible that a diet rich in fish oil may be of benefit [117]. In fact regular consumption of fresh, oily fish has been associated with a reduced risk of current asthma in childhood [121]. Furthermore children born to mothers with a history of asthma, had 80% lower odds of developing asthma when mothers ate oily fish at least monthly during pregnancy compared to no consumption. Nevertheless, fish consumption during pregnancy did not benefit children of non-asthmatic mothers. In contrast, fish stick consumption, rich in trans fats, during pregnancy increased 2-fold the odds of developing asthma [122]. In a recent study, fish consumption and high ratio of n-3 to n-6 polyunsaturated fatty acid intake has been independently associated with a decreased prevalence of asthma after age 18 and current asthma in young female Japanese adults. Moreover the inverse association between fish intake and asthma persisted after additional adjustment for docosahexaenoic acid intake indicating that possibly other constituents in fish might influence this association [123].

On the other hand there are studies that show either insignificant associations or inverse associations between asthma and fish consumption. The Nurses' Health Study, investigating the incidence of adult-onset asthma in 77,866 women aged 34-68 years, found no relationship between dietary intake of fatty acids and the incidence of doctor-diagnosed asthma over a 10-year period [117]. Additionally in a Japanese population which is known to have higher intake of fish and lower incidence of asthma, a significantly higher prevalence of asthma was noted among subjects who ate fish one to two times a week than among those who ate fish one to two times a month. The risk increased gradually with increasing frequency of fish intake, and the positive trend was statistically significant. The positive association between fish intake and asthma could be explained by the inhibition of the cyclooxygenase pathway and decreased production of bronchodilator prostaglandins, associated with diversion of arachidonic acid metabolites into the lipoxygenase cascade, provoking asthmatic attacks. The other possibility is that salt may be related to the connection between fish intake and asthma, as the Japanese regularly eat fish cooked with salt or salty seasoning [124].

6. Neurodevelopment

Observational studies show positive associations between fish intake and maternal DHA levels during pregnancy and behavioural attention scores, visual recognition memory and language comprehension in infancy [12]. Maternal intake of DHA is beneficial for early neurodevelopment as large amounts of DHA are necessary for brain development, for neural membrane production used for synaptogenesis, axonal and dendritic outgrowth, both in brain and retina [61]. DHA is incorporated into the rapidly developing brain during gestation and the first two years of infancy, concentrating in gray matter and retinal membranes [12]. Moreover an optimal balance in n-3 to n-6 long chain fatty acid ratio of 1:4 should be achieved for proper neurodevelopment and cognitive functions mainly due to improvement of membrane fluidity [125].

However, although maternal fish consumption is supposed to have beneficial effects on development of infants due to its high content in n-3 long chain polyunsaturated fatty acids, necessary for healthy development and function of the nervous system, it may be harmful for child cognitive development, since fish is a common source of methylmercury and other contaminants [60, 126, 127].

Methylmercury, a neurotoxin to which the fetal brain is particularly sensitive, is absorbed by the mother during pregnancy, it easily traverses the placenta and is amassed in the blood, brain, and other tissues of the fetus. Methylmercury has profound adverse effects on the immature central nervous system of the fetus and infant, as it inhibits the division, differentiation and migration of neuronal cells and disrupts the architecture of the developing brain [2, 82, 128]. There is an inverse and significant relationship observed between prenatal mercury exposure and both cognitive and psychomotor function among one-year-old infants. Methylmercury's prenatal exposure in infants and children is associated with serious health problems such as brain damage, mental retardation, cerebral palsy, deafness, blindness, seizures, lower neurologic optimality score, and neuropsychological deficits, dysarthria and poor language development, as well as sensory impairment and motor impairment [8, 127, 128]. The results of a recent study showed that adjusted cognitive deficit due to higher mercury exposure was significant among one-year-olds, but became insignificant at 24 and 36 months of age, suggesting that the cognitive and motor deficits possibly attributed to the prenatal higher mercury exposure level in one year-olds are subtle or easily reversible over the next two years of life. However, much more efforts must be dedicated to explain the real biological meaning of the mental and psychomotor deficits in infants at 12 months of age resulting possibly from prenatal mercury exposure [127].

There are also studies that show that consumption of a diet high in ocean fish does not appear to pose threat as far as developmental outcomes of young children are concerned [129-131]. In a US cohort with moderate levels of fish consumption, pregnant women who ate more fish had higher erythrocyte mercury levels. Among their children, higher prenatal mercury exposure was associated with lower developmental test scores at age 3 years. Nevertheless, no overall adverse effect upon child development was observed with higher maternal fish intake. Rather, maternal fish intake more than twice a week was associated with improved performance on tests of language and visual motor skills [131]. In addition no consistent evidence for adverse effects from prenatal methylmercury secondary to fish consumption was observed at the exposure levels found in the Seychelles cohort, however possible adverse effect in the uppermost range, above 12 ppm of prenatal exposure included in this cohort has been suggested [132]. It may be possible that the exposure-outcome relationship may not be homogeneous across all individuals, suggesting that the effects of prenatal methylmercury exposure from maternal fish consumption during pregnancy are complex between children with different backgrounds and developmental environments [130].

Furthermore, consumption of fatty fish, a known route of exposure to persistent organic pollutants, could be associated with reduced fetal growth. High intake of fatty fish was found to be inversely related to birth weight, birth length, and head circumference. For pregnant women, the type of fish consumed is important, and moderate consumption of fatty fish should be encouraged [4]. Additionally an association between in utero polychlorinated biphenyls exposure and impairments of memory during infancy and childhood has been observed. Recently a similar association has been proposed between polychlorinated biphenyls exposure from Great Lakes fish and short-term memory processing in infancy, as participants with serum polychlorinated biphenyls concentrations in the upper quartile scored, on average, about 2 points lower than those in the lowest quartile [133].

The overall effect of fish consumption during pregnancy, incorporating the risks as well as the benefits, remains uncertain. Nevertheless, it seems that, on balance, maternal fish intake is associated with improved child cognitive development. Encouraging maternal consumption of fish lower in mercury and reduced environmental mercury contamination, would offer stronger benefits of fish intake [126, 131]. Interventions intended to reduce exposure to low levels of mercury in early life must, however, be carefully evaluated in consideration of the potential attendant harm from resultant behavior changes, such as reduced docosahexaenoic acid exposure from lower seafood intake [126]. Therefore, rather than limiting fish intake, pregnant women should consume smaller-size fish and shellfish with low methylmercury content to meet the recommended amount of n-3 fatty acid and obtain other health benefits from fish.

7. Neuropsychiatric Disorders

Epidemiological observations on Japanese and Icelanders suggest that the lack of seasonal affective disorders may be related to the high fish content in the diet [134]. Moreover a growing body of evidence indicates that fish consumption may favorably affect clinical neurologic outcomes in adults, including cognitive decline, dementia, schizophrenia, major depressive disorder, post-partum depression, borderline personality disorder, bipolar disorder, hostility, anxiety disorders, and other neuropsychiatric disorders, although some recent studies with observational design have found contradictory results. Possible explanations for the inconsistency observed, apart from the complex nature of these varied psychiatric disease states and multifactorial genetic and epigenetic mechanisms, could be the mercury content of fish or reverse causality [12, 30, 71].

It is important to mention that essential fatty acids are major constituents of synaptic membranes and are believed to play an integral role in neurological function including signal transduction [30]. In fact, epidemiological evidence suggests that n-3 fatty acid levels in erythrocyte membranes of patients with central nervous system disorders such as Alzheimer's disease, schizophrenia, depression and attention-deficit hyperactivity disorder are significantly reduced. However it still is unknown whether these low levels are of primary aetiological significance or are secondary to the diseases [125].

7.1. Alzheimer's Disease

Low intakes of fish or n-3 polyunsaturated fatty acids have been associated with increased risk of Alzheimer's disease [135]. Fish and especially fatty fish consumption has been associated with a reduced risk of impaired cognitive function in middle-aged or elderly populations [125, 136-139] and postponement of the process of Alzheimer's disease [61].

In the elderly a diet high in fish and fish products has been associated with better cognitive performance in a dose-dependent manner which suggests that a higher intake of EPA and DHA is associated with less cognitive decline [138, 140]. According to a recent study, fish consumers had significantly less 5-year subsequent cognitive decline than did non fish consumers. Consumption of 6 servings of 140 gr lean fish per week or 1 serving of 140 gr fatty fish per week, was associated with less subsequent cognitive decline in elderly men [140]. Furthermore subjects whose mean daily intake of fish and fish products was more than 10 gr/day had significantly better mean test scores and a lower prevalence of poor cognitive performance than did those whose intake was less than 10 gr/day. The maximum effect was observed at an intake of approximately 75 gr/day. The effect was more pronounced for non-processed lean fish and fatty fish [138]. Persons who consumed at least one fish meal per week had 60% less risk of Alzheimer's disease than did persons who reported eating fish rarely or never [141]. According to another study, consumption of more than 2 servings of fish per week was associated with a 50% reduction in the risk of Alzheimer's disease [142]. Finally, fatty fish and marine n-3 PUFA consumption were significantly associated with a decreased risk of global cognitive function impairment and psychomotor speed by 19–28% [143].

Weekly consumption of fish may decrease the risk of dementia and Alzheimer's disease, especially among ApoE epsilon 4 non-carriers [144]. According to the Cardiovascular Health Cognition Study, consumption of fatty fish, but not lean fried fish, more than twice per week, was associated with a reduced risk of dementia by 28% (95%CI: 0.51-1.02) and Alzheimer's disease by 41% (95%CI: 0.36-0.95) in comparison to those who ate fish less than once per month, for those without the ApoE epsilon 4 allele [145].

The beneficial association between fish consumption and cognition is probably attributed to the n-3 fatty acids. These essential polyunsaturated fatty acids may play a role in the reduction of inflammatory processes in the brain and may have a specific role in brain development and regeneration of nerve cells [146, 147]. The classic b-amyloid deposits of Alzheimer's disease tend to coincide with blood vessels in the brain. Therefore, the inflammation caused by most fatty acids, which are the highest dietary risk factor for Alzheimer's disease, could be regulated by the n-3 fatty acids [148]. However it should be kept in mind that there might be an association between achievement of higher education and healthy dietary habits acquired early in life and that highly educated people might adhere more easily to dietary recommendations regarding fish consumption [146, 147].

7.2. Depression

Depressed mood has been associated with fish intake both directly and indirectly, as part of a diet or a lifestyle that is associated with depression [149]. Fish and seafood are the major source of n-3 fatty acids in the human diet, required for optimal neuronal function with shown efficacy as adjunctive treatment in several psychiatric disorders. Thus infrequent consumption of fish could mean a low intake of n-3 fatty acids and an elevated risk of depression [135, 150]. Low levels of n-3 fatty acids have been found in the red blood cells and plasma of depressed patients compared to controls, providing evidence for a negative association between depressed mood and dietary fish intake [149]. A possible mechanism linking fish consumption and depression may be the regulation of serotonergic nervous system function, as lower serotonergic activity has been well established in the pathophysiology of depression. High cerebrospinal fluid concentrations of 5-hydroxy-indoleacetic acid, a metabolite of serotonin, have been shown to be associated with high plasma concentrations of n-3 polyunsaturated fatty acids among healthy subjects [150].

An epidemiological study investigating fish consumption worldwide has demonstrated an increased prevalence of major depression in nations with less fish consumption per capita, indicating a negative correlation between fish consumption and worldwide depression prevalence. The correlation between apparent fish consumption and lower annual prevalence of major depression does not show that fish consumption can cause differences in the prevalence of major depression or that eating fish could be suggested as useful treatment for depression, since various cultural, economic, social, and other factors can confound this relation [125, 151]. In addition according to two cross-sectional studies based on the adult population from Finland, a higher risk of depressive symptoms among infrequent consumers of fish was observed, with the strength of the association being especially impressive among women [150, 152]. Low frequency of fish consumption was significantly associated with depression and suicidal thoughts in women, but not among men. In fact low frequency of fish consumption increased the risk about 2.5-fold for subsequent depression in females. The reason for the gender difference remains unclear but it could be explained by differences in neurotransmitter and phospholipid metabolism between sexes, suggesting an association between fish consumption, changes in serotonin metabolism and female gender [152].

7.3. Postpartum Depression

Both lower concentrations of docosahexaenoic acid in mothers' milk and lower national rates of seafood consumption have been robustly correlated with higher rates of major postpartum depressive symptoms. A nearly 50-fold difference in the prevalence rates of major postpartum depressive symptoms across countries has been observed and has been found to be substantially associated with n-3 fatty acid nutritional status. According to a cross-national ecological analysis, higher docosahexaenoic acid content in mother's milk and greater seafood intake were highly correlated with a lower prevalence of postpartum depression, indicating that mothers may be at a higher risk of suffering postpartum depression when they become depleted of n-3 essential fatty acids, in particular docosahexaenoic acid. Therefore inadequate dietary intake of n-3 fatty acids and the subsequent maternal depletion of n-3 fatty acids during pregnancy are associated with an increased risk of major postpartum depressive symptoms [153].

Nevertheless there are also studies that failed to demonstrate a clear inverse relationship between fish and n-3 polyunsaturated fatty acid intake and postpartum depression. Prenatal fish consumption was not predictive of postnatal depression, and postnatal n-3 status was not associated with postnatal depression. However, prenatal fish consumption did predict n-3 status after birth [154]. In addition consumption of n-3 polyunsaturated fatty acids, especially docosahexaenoic acid, in the third quartile was marginally independently associated with a decreased risk of postpartum depression, although the linear trends were not statistically significant [155].

8. Fish Contaminants

Fish and shellfish supply the human diet with not only complex nutrients including the n-3 fatty acids, but also highly toxic chemicals [83], some of them, such as heavy metals and organohalogenated compounds, with well known adverse effects on human health [156]. Among metals, methylmercury is the compound of most concern, but fish may also contain arsenic, cadmium, lead, and organotin compounds. Among organohalogenated contaminants, most notable are polychlorinated dioxins and furans, polychlorinated biphenyls, various brominated flame retardants such as polybrominated diphenyl ethers, and various organochlorine insecticides [7, 12, 13, 83, 157, 158].

8.1. Mercury

Mercury, a reactive heavy metal, is emitted from natural and human sources. From the atmosphere, mercury cycles from rainwater into lakes and oceans, where it is converted by microbial activity into organic methylmercury. Despite the fact that inorganic mercury is poorly absorbed following ingestion and elemental mercury does not readily cross tissue barriers, methylmercury is readily absorbed and actively transported into tissues. It is known that methylmercury crosses the blood–brain barriers and that the highest levels are found in the kidneys. Methylmercury is slowly transformed into other forms of mercury and is excreted mainly in feces, urine, and breast milk, mostly within a few weeks to a few months after exposure [8, 12]. Mercury can accumulate in the body if eaten at a greater rates than excreted [159]. However, while it is generally believed that passage out of the human body takes three months to a year for mercury, it may have a different half-life in different organs, lingering longer in the brain and kidneys than in the blood. Genetic susceptibility and synergy with other compounds in the body may play roles in determining at what threshold these contaminants prove dangerous to individuals [8].

Mercury can cause several adverse health effects, including impairment of the immune, cardiovascular, and reproductive systems [159]. At high doses, methylmercury has been well established as a neurotoxic agent for humans and many species of animals [83]. Health effects of very high mercury exposure include paresthesias, ataxia, and sensory abnormalities in adults [12]. Methylmercury crosses the placenta, and as fetal exposure correlates with maternal exposure, marked neurodevelopmental abnormalities and delayed cognitive and neuromuscular development have been observed in children following very high gestational exposure [12, 13]. Finally it has been suggested that the cardiovascular system may be adversely affected by lower methylmercury exposures than those currently associated with neurological and neurodevelopmental deficits [83].

The distribution of mercury among different species is in accordance with the process of mercury uptake by fish and the interaction of biotic parameters such as size, sex, longevity, growth rate, feeding habits, trophic position and habitat [160]. Larger and longer-living predators, such as swordfish, tilefish tuna, orange roughy and shark have higher tissue concentrations, while smaller or shorter-lived species that are not predatory like sardines or salmon have very low concentrations [8, 12, 13, 83, 161]. It is important to mention that farmed fish have the lowest levels of methyl mercury [162]. In addition methylmercury in fish meat is bound to muscle proteins, therefore skimming, trimming of fish, or cooking does not significantly reduce its concentration [1, 13, 16, 128, 163].

8.2. Polychlorinated Biphenyls

Polychlorinated biphenyls (PCBs), dibenzo-p-dioxins and dibenzo-furans are toxic and persistent environmental synthetic pollutants, previously used in industrial and commercial processes, mainly generated as undesirable byproducts during waste incineration, metal refinery, production of chlorinated herbicides and bleaching of paper [12, 164]. Cancer risk is the greatest toxicological concern for all dioxins and dioxin-like polychlorinated biphenyls. In particular the most potent dioxin, TCDD (2,3,7,8-tetrachlorodibenzo-*p*-dioxin) is classified as a group I carcinogen, thus carcinogenic to human [13]. Other carcinogenic compounds present in fish are DDT, dieldrin and heptachlor [12].

Like mercury, polychlorinated biphenyls are also dangerous to adults as well as unborn and very young children [8]. Adverse effects of these compounds include immunotoxicity, reproductive toxicity, endocrine disruption, and developmental neurotoxicity. Dioxin exposure may also be associated with an increased risk of diabetes [12, 13, 164]. Furthermore polychlorinated biphenyls' exposure during adulthood has been associated with increased risk of neuropsychological impairment, impairments in memory and learning, whereas executive and visual memory and executive and visual–spatial function were unaffected [133].

As these compounds are lipophilic, high levels are found in fatty fish such as salmon and herring, whereas lean ocean fish such as cod, flounder, and haddock are the least likely to be contaminated. Farmed and wild-caught fish appear to have similar levels of contaminants [156]. As with mercury, the only treatment for polychlorinated biphenyls poisoning is removal of the source. However, removing the skin and fatty areas, such as the belly, back, and dark side meat, as well as trimming the fat from fish bodies before cooking, has been shown to reduce levels of these contaminants in fish [8, 13, 16]. Their exposure can be further minimized by certain cooking procedures, such as using grills in cooking or cooking the fish on a rack to allow fat to drain [1, 163]. However frying fatty fish is ill-advised [8].

It is apparent that there are benefits to eating fish, and to some extent these benefits derive from their fatty acid content. Therefore choosing fish low in methylmercury and polychlorinated biphenyls and high in n-3 fatty acids is desirable [165]. Additionally it is important to mention that although there are many species of fish that are rich sources of the n-3 fatty acids and have lower methylmercury concentrations, these species may not be consistently low in other organic contaminants, such as dioxins and polychlorinated biphenyls. [83].

Dietary Guidelines

According to the American Heart Association dietary guidelines, healthy adults should eat at least two servings of fish and particularly fatty fish per week. In addition patients with or without documented cardiovascular disease should consume a variety of fish at least twice a week. However, the Scientific Statement of the American Heart Association also indicates that fish and seafood is potentially a major source of exposure to various environmental contaminants such as methyl mercury and PCBs. Depending on their stage of life, consumers need to be aware of the potential health risks of eating fish [16, 163].

Fish with higher mercury concentrations should be consumed in smaller quantities. There are fish species that are low in mercury and rich sources of n-3 fatty acids, while other fish species are highly contaminated with mercury but not especially rich sources of n-3 fatty acids. Therefore avoidance of the fish species higher in methylmercury will not compromise nutritional intake of n-3 fatty acids from fish and shellfish [83, 166]. Especially for individuals who consume more than 5 servings per week, they should limit the intake of species highest in mercury levels. As far as levels of dioxins and polychlorinated biphenyls in fish are concerned, potential carcinogenic or other effects are possibly outweighed by the potential benefits of fish intake and should have little impact on choices or consumption of seafood [12, 167].

The fish species rich in n-3 PUFAs are sardines, mackerel, herring, trout, salmon, anchovy, sablefish and bluefish. They provide an adequate amount of n-3 PUFAs (2.7–7.5 g per meal) and appear to meet the nutritional recommendations of the American Heart Association [1]. Ocean fish that have been reported to have low methylmercury levels or average levels below 0.4 mg/kg are black drum, butterfish, canned light tuna, cod, crab, croaker, flounder, haddock, halibut, herring, jacksmelt, southern kingfish, lobster, mahi-mahi, ocean perch, oysters, pollack, pompano, red drum, scallops, shrimp, speckled trout, sheepshead, skate, spot, tripletail, whitefish, and white grunt. Freshwater fish that have been reported to have low methylmercury levels or average levels below 0.4 mg/kg are bluegill sunfish, farm-raised catfish, trout, and crayfish, salmon, tilapia and trout [161].

Guidelines for Women of Childbearing Age and Children

As far as women of childbearing age, nursing mothers and very young children are concerned, they should consume two seafood servings per week, restrict the consumption of certain species which are high in methyl-mercury, such as shark, swordfish, golden bass, and king mackerel to 1 meal per 2 weeks and consult regional advisories for locally caught freshwater fish. Consumption of 2–3 meals of other fish per week like sardines, salmon, or shrimp is advised. In non-pregnant individuals the recommended dietary fish intake is 1 and 2–3 per week, respectively, for high and low methyl-mercury-containing fish [10, 13, 16, 82, 127, 163, 167]. Recommendations for fish consumption during pregnancy should take into account the nutritional benefits of fish, as well as the potential harms from mercury exposure [131]. Emphasis must placed on adequate consumption of other fish and shellfish in order to provide reasonable amounts of DHA and avoid further decreases in already low seafood intake among women. Low fish consumption due to confusion regarding risks and benefits could result in suboptimal neurodevelopment in children [12].

Conclusion

Fish consumption is an important part of a healthy diet when consumed in moderation and when a variety of species is selected. The type of fish, the frequency of consumption and the meal size are essential issues in the balance of risks and benefits of regular fish consumption [7]. Because different types of fish have different concentrations of both n-3 fatty acids and chemical contaminants, knowledge of how each component individually influences risk is of considerable interest [168].

References

[1] Sidhu KS. Health benefits and potential risks related to consumption of fish or fish oil. *Regul. Toxicol. Pharmacol.* 2003;38:336-44.

[2] Smith KM, Sahyoun NR. Fish consumption: recommendations versus advisories, can they be reconciled? Nutr Rev 2005;63:39-46.

[3] Engeset D, Alsaker E, Lund E, et al. Fish consumption and breast cancer risk. The European Prospective Investigation into Cancer and Nutrition (EPIC). *Int. J. Cancer.* 2006;119:175-82.

[4] Halldorsson TI, Meltzer HM, Thorsdottir I, Knudsen V, Olsen SF. Is high consumption of fatty fish during pregnancy a risk factor for fetal growth retardation? A study of 44,824 Danish pregnant women. *Am. J. Epidemiol.* 2007;166:687-96.

[5] Schmidt EB, Rasmussen LH, Rasmussen JG, Joensen AM, Madsen MB, Christensen JH. Fish, marine n-3 polyunsaturated fatty acids and coronary heart disease: a minireview with focus on clinical trial data. Prostaglandins Leukot Essent Fatty Acids 2006;75:191-5.

[6] Schmidt EB, Skou HA, Christensen JH, Dyerberg J. N-3 fatty acids from fish and coronary artery disease: implications for public health. Public Health Nutr 2000;3:91-8.

[7] Domingo JL. Omega-3 fatty acids and the benefits of fish consumption: is all that glitters gold? *Environ. Int.* 2007;33:993-8.

[8] Wilson JF. Balancing the risks and benefits of fish consumption. *Ann. Intern. Med.* 2004;141:977-80.

[9] Zhang J, Sasaki S, Amano K, Kesteloot H. Fish consumption and mortality from all causes, ischemic heart disease, and stroke: an ecological study. *Prev. Med.* 1999;28:520-9.

[10] Arnold SM, Lynn TV, Verbrugge LA, Middaugh JP. Human biomonitoring to optimize fish consumption advice: reducing uncertainty when evaluating benefits and risks. *Am. J. Public Health* 2005;95:393-7.

[11] Fernandez E, Chatenoud L, La Vecchia C, Negri E, Franceschi S. Fish consumption and cancer risk. *Am. J. Clin. Nutr* .1999;70:85-90.

[12] Mozaffarian D, Rimm EB. Fish intake, contaminants, and human health: evaluating the risks and the benefits. *Jama* 2006;296:1885-99.

[13] Costa LG. Contaminants in fish: risk-benefit considerations. *Arh. Hig. Rada Toksikol.* 2007;58:367-74.

[14] Gillum RF, Mussolino M, Madans JH. The relation between fish consumption, death from all causes, and incidence of coronary heart disease. the NHANES I Epidemiologic Follow-up Study. *J. Clin. Epidemiol.* 2000;53:237-44.

[15] Wang C, Harris WS, Chung M, et al. n-3 Fatty acids from fish or fish-oil supplements, but not alpha-linolenic acid, benefit cardiovascular disease outcomes in primary- and secondary-prevention studies: a systematic review. *Am. J. Clin. Nutr* .2006;84:5-17.

[16] Kris-Etherton PM, Harris WS, Appel LJ. Fish consumption, fish oil, omega-3 fatty acids, and cardiovascular disease. *Arterioscler. Thromb. Vasc. Biol.* 2003;23:e20-30.

[17] Burr ML, Fehily AM, Gilbert JF, et al. Effects of changes in fat, fish, and fibre intakes on death and myocardial reinfarction: diet and reinfarction trial (DART). *Lancet* 1989;2:757-61.

[18] Burr ML, Sweetham PM, Fehily AM. Diet and reinfarction. *Eur. Heart J.* 1994;15:1152-3.

[19] Bang HO, Dyerberg J, Sinclair HM. The composition of the Eskimo food in north western Greenland. *Am. J. Clin. Nutr.* 1980;33:2657-61.

[20] Bang HO, Dyerberg J, Hjoorne N. The composition of food consumed by Greenland Eskimos. *Acta Med. Scand.* 1976;200:69-73.

[21] Daviglus ML, Stamler J, Greenland P, Dyer AR, Liu K. Fish consumption and risk of coronary heart disease. What does the evidence show? *Eur. Heart J.* 1997;18:1841-2.

[22] Mozaffarian D. Fish, n-3 fatty acids, and cardiovascular haemodynamics. J Cardiovasc Med (Hagerstown) 2007;8 Suppl 1:S23-6.

[23] He K, Song Y, Daviglus ML, et al. Accumulated evidence on fish consumption and coronary heart disease mortality: a meta-analysis of cohort studies. *Circulation* 2004;109:2705-11.

[24] Hu FB, Bronner L, Willett WC, et al. Fish and omega-3 fatty acid intake and risk of coronary heart disease in women. *Jama* 2002;287:1815-21.

[25] Oomen CM, Feskens EJ, Rasanen L, et al. Fish consumption and coronary heart disease mortality in Finland, Italy, and The Netherlands. *Am. J. Epidemiol.* 2000;151:999-1006.

[26] Konig A, Bouzan C, Cohen JT, et al. A quantitative analysis of fish consumption and coronary heart disease mortality. *Am. J. Prev. Med.* 2005;29:335-46.

[27] Whelton SP, He J, Whelton PK, Muntner P. Meta-analysis of observational studies on fish intake and coronary heart disease. *Am. J. Cardiol.* 2004;93:1119-23.

[28] Daviglus ML, Stamler J, Orencia AJ, et al. Fish consumption and the 30-year risk of fatal myocardial infarction. *N. Engl. J .Med.* 1997;336:1046-53.

[29] Yuan JM, Ross RK, Gao YT, Yu MC. Fish and shellfish consumption in relation to death from myocardial infarction among men in Shanghai, China. *Am. J. Epidemiol.* 2001;154:809-16.

[30] Holub DJ, Holub BJ. Omega-3 fatty acids from fish oils and cardiovascular disease. *Mol. Cell Biochem.* 2004;263:217-25.

[31] Erkkila AT, Lehto S, Pyorala K, Uusitupa MI. n-3 Fatty acids and 5-y risks of death and cardiovascular disease events in patients with coronary artery disease. *Am. J. Clin. Nutr.* 2003;78:65-71.

[32] Mozaffarian D, Lemaitre RN, Kuller LH, Burke GL, Tracy RP, Siscovick DS. Cardiac benefits of fish consumption may depend on the type of fish meal consumed: the Cardiovascular Health Study. *Circulation* 2003;107:1372-7.

[33] Connor WE. Will the dietary intake of fish prevent atherosclerosis in diabetic women? *Am. J. Clin. Nutr.* 2004;80:535-6.

[34] Iso H, Kobayashi M, Ishihara J, et al. Intake of fish and n3 fatty acids and risk of coronary heart disease among Japanese: the Japan Public Health Center-Based (JPHC) Study Cohort I. *Circulation* 2006;113:195-202.

[35] Kromhout D, Feskens EJ, Bowles CH. The protective effect of a small amount of fish on coronary heart disease mortality in an elderly population. *Int. J. Epidemiol.* 1995;24:340-5.

[36] Panagiotakos DB, Pitsavos C, Zampelas A, et al. Fish consumption and the risk of developing acute coronary syndromes: the CARDIO2000 study. *Int. J. Cardiol.* 2005;102:403-9.

[37] Panagiotakos DB, Pitsavos C, Zampelas A, Chrysohoou C, Stefanadis C. The relationship between fish consumption and the risk of developing acute coronary syndromes among smokers: the CARDIO2000 case-control study. *Nutr. Metab. Cardiovasc. Dis.* 2005;15:402-9.

[38] Rodriguez BL, Sharp DS, Abbott RD, et al. Fish intake may limit the increase in risk of coronary heart disease morbidity and mortality among heavy smokers. The Honolulu Heart Program. *Circulation* 1996;94:952-6.

[39] Hallgren CG, Hallmans G, Jansson JH, et al. Markers of high fish intake are associated with decreased risk of a first myocardial infarction. *Br. J. Nutr* .2001;86:397-404.

[40] Tavani A, Pelucchi C, Negri E, Bertuzzi M, La Vecchia C. n-3 Polyunsaturated fatty acids, fish, and nonfatal acute myocardial infarction. *Circulation* 2001;104:2269-72.

[41] Landmark K, Abdelnoor M, Kilhovd B, Dorum HP. Eating fish may reduce infarct size and the occurrence of Q wave infarcts. *Eur. J. Clin. Nutr.* 1998;52:40-4.

[42] Hu FB, Cho E, Rexrode KM, Albert CM, Manson JE. Fish and long-chain omega-3 fatty acid intake and risk of coronary heart disease and total mortality in diabetic women. *Circulation* 2003;107:1852-7.

[43] Siscovick DS, Raghunathan TE, King I, et al. Dietary intake and cell membrane levels of long-chain n-3 polyunsaturated fatty acids and the risk of primary cardiac arrest. *Jama* 1995;274:1363-7.

[44] Chrysohoou C, Panagiotakos DB, Pitsavos C, et al. Long-term fish consumption is associated with protection against arrhythmia in healthy persons in a Mediterranean region--the ATTICA study. *Am. J. Clin. Nutr.* 2007;85:1385-91.

[45] Albert CM, Hennekens CH, O'Donnell CJ, et al. Fish consumption and risk of sudden cardiac death. *Jama* 1998;279:23-8.

[46] Bouzan C, Cohen JT, Connor WE, et al. A quantitative analysis of fish consumption and stroke risk. *Am. J. Prev. Med.* 2005;29:347-52.

[47] Skerrett PJ, Hennekens CH. Consumption of fish and fish oils and decreased risk of stroke. *Prev. Cardiol.* 2003;6:38-41.

[48] Gillum RF, Mussolino ME, Madans JH. The relationship between fish consumption and stroke incidence. The NHANES I Epidemiologic Follow-up Study (National Health and Nutrition Examination Survey). *Arch. Intern. Med.* 1996;156:537-42.

[49] Keli SO, Feskens EJ, Kromhout D. Fish consumption and risk of stroke. The Zutphen Study. *Stroke* 1994;25:328-32.

[50] He K, Rimm EB, Merchant A, et al. Fish consumption and risk of stroke in men. *Jama* 2002;288:3130-6.

[51] He K, Song Y, Daviglus ML, et al. Fish consumption and incidence of stroke: a meta-analysis of cohort studies. *Stroke* 2004;35:1538-42.

[52] Iso H, Rexrode KM, Stampfer MJ, et al. Intake of fish and omega-3 fatty acids and risk of stroke in women. *Jama* 2001;285:304-12.

[53] Mozaffarian D, Longstreth WT, Jr., Lemaitre RN, et al. Fish consumption and stroke risk in elderly individuals: the cardiovascular health study. *Arch. Intern. Med.* 2005;165:200-6.

[54] Orencia AJ, Daviglus ML, Dyer AR, Shekelle RB, Stamler J. Fish consumption and stroke in men. 30-year findings of the Chicago Western Electric Study. *Stroke* 1996;27:204-9.

[55] Caicoya M. Fish consumption and stroke: a community case-control study in Asturias, Spain. *Neuroepidemiology* 2002;21:107-14.

[56] Morris MC, Manson JE, Rosner B, Buring JE, Willett WC, Hennekens CH. Fish consumption and cardiovascular disease in the physicians' health study: a prospective study. *Am. J. Epidemiol.* 1995;142:166-75.

[57] Myint PK, Welch AA, Bingham SA, et al. Habitual fish consumption and risk of incident stroke: the European Prospective Investigation into Cancer (EPIC)-Norfolk prospective population study. *Public Health Nutr.* 2006;9:882-8.

[58] Kaushik S, Wang JJ, Flood V, Liew G, Smith W, Mitchell P. Frequency of fish consumption, retinal microvascular signs and vascular mortality. *Microcirculation* 2008;15:27-36.

[59] Wennberg M, Bergdahl IA, Stegmayr B, et al. Fish intake, mercury, long-chain n-3 polyunsaturated fatty acids and risk of stroke in northern Sweden. *Br. J. Nutr.* 2007;98:1038-45.

[60] McMichael AJ, Butler CD. Fish, health, and sustainability. *Am. J. Prev. Med.* 2005;29:322-3.

[61] Hooijmans CR, Kiliaan AJ. Fatty acids, lipid metabolism and Alzheimer pathology. *Eur. J. Pharmacol.* 2008;585:176-96.

[62] Panagiotakos DB, Zeimbekis A, Boutziouka V, et al. Long-term fish intake is associated with better lipid profile, arterial blood pressure, and blood glucose levels in elderly people from Mediterranean islands (MEDIS epidemiological study). *Med. Sci. Monit* 2007;13:CR307-12.

[63] Moore CS, Bryant SP, Mishra GD, et al. Oily fish reduces plasma triacylglycerols: a primary prevention study in overweight men and women. *Nutrition* 2006;22:1012-24.

[64] Das UN. Beneficial effect(s) of n-3 fatty acids in cardiovascular diseases: but, why and how? Prostaglandins Leukot Essent Fatty Acids 2000;63:351-62.

[65] Nkondjock A, Receveur O. Fish-seafood consumption, obesity, and risk of type 2 diabetes: an ecological study. *Diabetes Metab.* 2003;29:635-42.

[66] Nakamura Y, Ueno Y, Tamaki S, et al. Fish consumption and early atherosclerosis in middle-aged men. *Metabolism* 2007;56:1060-4.

[67] Bao DQ, Mori TA, Burke V, Puddey IB, Beilin LJ. Effects of dietary fish and weight reduction on ambulatory blood pressure in overweight hypertensives. *Hypertension* 1998;32:710-7.

[68] Mori TA, Burke V, Puddey IB, Shaw JE, Beilin LJ. Effect of fish diets and weight loss on serum leptin concentration in overweight, treated-hypertensive subjects. *J. Hypertens.* 2004;22:1983-90.

[69] Mori TA, Bao DQ, Burke V, Puddey IB, Watts GF, Beilin LJ. Dietary fish as a major component of a weight-loss diet: effect on serum lipids, glucose, and insulin metabolism in overweight hypertensive subjects. *Am. J. Clin. Nutr.* 1999;70:817-25.

[70] Erkkila AT, Lichtenstein AH, Mozaffarian D, Herrington DM. Fish intake is associated with a reduced progression of coronary artery atherosclerosis in postmenopausal women with coronary artery disease. *Am. J. Clin. Nutr.* 2004;80:626-32.

[71] Sanchez-Villegas A, Henriquez P, Figueiras A, Ortuno F, Lahortiga F, Martinez-Gonzalez MA. Long chain omega-3 fatty acids intake, fish consumption and mental disorders in the SUN cohort study. *Eur. J. Nutr.* 2007;46:337-46.

[72] Carpentier YA, Portois L, Malaisse WJ. n-3 fatty acids and the metabolic syndrome. *Am. J. Clin. Nutr.* 2006;83:1499S-1504S.

[73] Zampelas A, Panagiotakos DB, Pitsavos C, et al. Fish consumption among healthy adults is associated with decreased levels of inflammatory markers related to cardiovascular disease: the ATTICA study. *J .Am. Coll Cardiol.* 2005;46:120-4.

[74] De Lorgeril M, Salen P. Fish and N-3 fatty acids for the prevention and treatment of coronary heart disease: nutrition is not pharmacology. *Am. J. Med.* 2002;112:316-9.

[75] Marckmann P, Gronbaek M. Fish consumption and coronary heart disease mortality. A systematic review of prospective cohort studies. *Eur. J. Clin. Nutr.* 1999;53:585-90.

[76] Osler M, Andreasen AH, Hoidrup S. No inverse association between fish consumption and risk of death from all-causes, and incidence of coronary heart disease in middle-aged, Danish adults. *J. Clin. Epidemiol.* 2003;56:274-9.

[77] Ascherio A, Rimm EB, Stampfer MJ, Giovannucci EL, Willett WC. Dietary intake of marine n-3 fatty acids, fish intake, and the risk of coronary disease among men. *N. Engl. J. Med.* 1995;332:977-82.

[78] Kromhout D, Bloemberg BP, Feskens EJ, Hertog MG, Menotti A, Blackburn H. Alcohol, fish, fibre and antioxidant vitamins intake do not explain population differences in coronary heart disease mortality. *Int. J.Epidemiol*.1996;25:753-9.

[79] Guallar E, Aro A, Jimenez FJ, et al. Omega-3 fatty acids in adipose tissue and risk of myocardial infarction: the EURAMIC study. Arterioscler *Thromb. Vasc. Biol.* 1999;19:1111-8.

[80] Sasazuki S. Case-control study of nonfatal myocardial infarction in relation to selected foods in Japanese men and women. *Jpn. Circ. J.* 2001;65:200-6.

[81] Burr ML, Ashfield-Watt PA, Dunstan FD, et al. Lack of benefit of dietary advice to men with angina: results of a controlled trial. *Eur. J. Clin. Nutr.* 2003;57:193-200.

[82] Levenson CW, Axelrad DM. Too much of a good thing? Update on fish consumption and mercury exposure. *Nutr. Rev.* 2006;64:139-45.
[83] Mahaffey KR. Fish and shellfish as dietary sources of methylmercury and the omega-3 fatty acids, eicosahexaenoic acid and docosahexaenoic acid: risks and benefits. *Environ. Res.* 2004;95:414-28.
[84] Salonen JT, Seppanen K, Nyyssonen K, et al. Intake of mercury from fish, lipid peroxidation, and the risk of myocardial infarction and coronary, cardiovascular, and any death in eastern Finnish men. *Circulation* 1995;91:645-55.
[85] Chan HM, Egeland GM. Fish consumption, mercury exposure, and heart diseases. *Nutr. Rev.* 2004;62:68-72.
[86] Yoshizawa K, Rimm EB, Morris JS, et al. Mercury and the risk of coronary heart disease in men. *N. Engl .J .Med.* 2002;347:1755-60.
[87] Virtanen JK, Voutilainen S, Rissanen TH, et al. Mercury, fish oils, and risk of acute coronary events and cardiovascular disease, coronary heart disease, and all-cause mortality in men in eastern Finland. *Arterioscler. Thromb Vasc. Biol.* 2005;25:228-33.
[88] Diet, nutrition, and the prevention of chronic diseases. Report of a joint WHO/FAO expert consultation,916 ed. Geneva: World Health Organization, 2003.
[89] Geelen A, Schouten JM, Kamphuis C, et al. Fish consumption, n-3 fatty acids, and colorectal cancer: a meta-analysis of prospective cohort studies. *Am. J. Epidemiol.* 2007;166:1116-25.
[90] Kimura Y, Kono S, Toyomura K, et al. Meat, fish and fat intake in relation to subsite-specific risk of colorectal cancer: The Fukuoka Colorectal Cancer Study. *Cancer Sci.* 2007;98:590-7.
[91] Norat T, Bingham S, Ferrari P, et al. Meat, fish, and colorectal cancer risk: the European Prospective Investigation into cancer and nutrition. *J. Natl. Cancer Inst.* 2005;97:906-16.
[92] Luchtenborg M, Weijenberg MP, de Goeij AF, et al. Meat and fish consumption, APC gene mutations and hMLH1 expression in colon and rectal cancer: a prospective cohort study (The Netherlands). *Cancer Causes Control.* 2005;16:1041-54.
[93] Engeset D, Andersen V, Hjartaker A, Lund E. Consumption of fish and risk of colon cancer in the Norwegian Women and Cancer (NOWAC) study. *Br. J. Nutr.* 2007;98:576-82.
[94] English DR, MacInnis RJ, Hodge AM, Hopper JL, Haydon AM, Giles GG. Red meat, chicken, and fish consumption and risk of colorectal cancer. *Cancer Epidemiol.* Biomarkers Prev 2004;13:1509-14.
[95] Kaizer L, Boyd NF, Kriukov V, Tritchler D. Fish consumption and breast cancer risk: an ecological study. *Nutr .Cancer* 1989;12:61-8.
[96] Gago-Dominguez M, Yuan JM, Sun CL, Lee HP, Yu MC. Opposing effects of dietary n-3 and n-6 fatty acids on mammary carcinogenesis: The Singapore Chinese Health Study. *Br. J. Cancer.* 2003;89:1686-92.
[97] Hirose K, Takezaki T, Hamajima N, Miura S, Tajima K. Dietary factors protective against breast cancer in Japanese premenopausal and postmenopausal women. *Int. J. Cancer* 2003;107:276-82.

[98] Terry P, Rohan TE, Wolk A, Maehle-Schmidt M, Magnusson C. Fish consumption and breast cancer risk. *Nutr .Cancer* 2002;44:1-6.

[99] Vatten LJ, Solvoll K, Loken EB. Frequency of meat and fish intake and risk of breast cancer in a prospective study of 14,500 Norwegian women. *Int. J. Cancer* 1990;46:12-5.

[100] Stripp C, Overvad K, Christensen J, et al. Fish intake is positively associated with breast cancer incidence rate. *J. Nutr.* 2003;133:3664-9.

[101] Terry PD, Rohan TE, Wolk A. Intakes of fish and marine fatty acids and the risks of cancers of the breast and prostate and of other hormone-related cancers: a review of the epidemiologic evidence. *Am. J .Clin. Nutr* .2003;77:532-43.

[102] Holmes MD, Hunter DJ, Colditz GA, et al. Association of dietary intake of fat and fatty acids with risk of breast cancer. *Jama* 1999;281:914-20.

[103] Terry P, Wolk A, Vainio H, Weiderpass E. Fatty fish consumption lowers the risk of endometrial cancer: a nationwide case-control study in Sweden. *Cancer Epidemiol. Biomarkers Prev* .2002;11:143-5.

[104] Augustsson K, Michaud DS, Rimm EB, et al. A prospective study of intake of fish and marine fatty acids and prostate cancer. *Cancer Epidemiol. Biomarkers Prev.* 2003;12:64-7.

[105] Terry P, Lichtenstein P, Feychting M, Ahlbom A, Wolk A. Fatty fish consumption and risk of prostate cancer. *Lancet* 2001;357:1764-6.

[106] Sato F, Shimazu T, Kuriyama S, et al. [Fish intake and the risk of prostate cancer in Japan: a prospective cohort study]. Nippon Hinyokika Gakkai Zasshi 2008;99:14-21.

[107] Hedelin M, Chang ET, Wiklund F, et al. Association of frequent consumption of fatty fish with prostate cancer risk is modified by COX-2 polymorphism. *Int. J. Cancer* 2007;120:398-405.

[108] Takezaki T, Inoue M, Kataoka H, et al. Diet and lung cancer risk from a 14-year population-based prospective study in Japan: with special reference to fish consumption. *Nutr. Cancer*. 2003;45:160-7.

[109] Marchand JL, Luce D, Goldberg P, Bugel I, Salomon C, Goldberg M. Dietary factors and the risk of lung cancer in New Caledonia (South Pacific). *Nutr .Cancer* 2002;42:18-24.

[110] Nyberg F, Agrenius V, Svartengren K, Svensson C, Pershagen G. Dietary factors and risk of lung cancer in never-smokers. *Int. J .Cancer* 1998;78:430-6.

[111] Dosil-Diaz O, Ruano-Ravina A, Gestal-Otero JJ, Barros-Dios JM. Meat and fish consumption and risk of lung cancer: A case-control study in Galicia, Spain. *Cancer* Lett 2007;252:115-22.

[112] Zhang J, Temme EH, Kesteloot H. Fish consumption is inversely associated with male lung cancer mortality in countries with high levels of cigarette smoking or animal fat consumption. *Int. J. Epidemiol* .2000;29:615-21.

[113] Takezaki T, Hirose K, Inoue M, et al. Dietary factors and lung cancer risk in Japanese: with special reference to fish consumption and adenocarcinomas. *Br. J. Cancer* 2001;84:1199-206.

[114] Wolk A, Larsson SC, Johansson JE, Ekman P. Long-term fatty fish consumption and renal cell carcinoma incidence in women. *Jama* 2006;296:1371-6.

[115] Horrobin DF. Low prevalences of coronary heart disease (CHD), psoriasis, asthma and rheumatoid arthritis in Eskimos: arc they caused by high dietary intake of eicosapentaenoic acid (EPA), a genetic variation of essential fatty acid (EFA) metabolism or a combination of both? *Med. Hypotheses* 1987;22:421-8.

[116] Stamp LK, James MJ, Cleland LG. Diet and rheumatoid arthritis: a review of the literature. *Semin. Arthritis Rheum.* 2005;35:77-94.

[117] Thien FC, Woods RK, Walters EH. Oily fish and asthma--a fishy story? Further studies are required before claims can be made of a beneficial effect of oily fish consumption on asthma. *Med. J. Aust*.1996;164:135-6.

[118] Shapiro JA, Koepsell TD, Voigt LF, Dugowson CE, Kestin M, Nelson JL. Diet and rheumatoid arthritis in women: a possible protective effect of fish consumption. *Epidemiology* 1996;7:256-63.

[119] Schwartz J, Weiss ST. The relationship of dietary fish intake to level of pulmonary function in the first National Health and Nutrition Survey (NHANES I). *Eur. Respir. J.* 1994;7:1821-4.

[120] Shahar E, Folsom AR, Melnick SL, et al. Dietary n-3 polyunsaturated fatty acids and smoking-related chronic obstructive pulmonary disease. Atherosclerosis Risk in Communities Study Investigators. *N. Engl. J. Med.* 1994;331:228-33.

[121] Hodge L, Salome CM, Peat JK, Haby MM, Xuan W, Woolcock AJ. Consumption of oily fish and childhood asthma risk. *Med. J. Aust.* 1996;164:137-40.

[122] Salam MT, Li YF, Langholz B, Gilliland FD. Maternal fish consumption during pregnancy and risk of early childhood asthma. *J. Asthma* 2005;42:513-8.

[123] Miyamoto S, Miyake Y, Sasaki S, et al. Fat and fish intake and asthma in Japanese women: baseline data from the Osaka Maternal and Child Health Study. *Int. J. Tuberc. Lung Dis.* 2007;11:103-9.

[124] Takemura Y, Sakurai Y, Honjo S, et al. The relationship between fish intake and the prevalence of asthma: the Tokorozawa childhood asthma and pollinosis study. *Prev. Med.* 2002;34:221-5.

[125] Assisi A, Banzi R, Buonocore C, et al. Fish oil and mental health: the role of n-3 long-chain polyunsaturated fatty acids in cognitive development and neurological disorders. *Int. Clin. Psychopharmacol.* 2006;21:319-36.

[126] Oken E, Bellinger DC. Fish consumption, methylmercury and child neurodevelopment. *Curr .Opin. Pediatr.* 2008;20:178-83.

[127] Jedrychowski W, Perera F, Jankowski J, et al. Fish consumption in pregnancy, cord blood mercury level and cognitive and psychomotor development of infants followed over the first three years of life: Krakow epidemiologic study. *Environ. Int.* 2007;33:1057-62.

[128] Park S, Johnson MA. Awareness of fish advisories and mercury exposure in women of childbearing age. *Nutr. Rev*.2006;64:250-6.

[129] Davidson PW, Myers GJ, Cox C, et al. Effects of prenatal and postnatal methylmercury exposure from fish consumption on neurodevelopment: outcomes at 66 months of age in the Seychelles Child Development Study. *Jama* 1998;280:701-7.

[130] Huang LS, Myers GJ, Davidson PW, et al. Is susceptibility to prenatal methylmercury exposure from fish consumption non-homogeneous? Tree-structured analysis for the Seychelles Child Development Study. Neurotoxicology 2007;28:1237-44.
[131] Oken E, Radesky JS, Wright RO, et al. Maternal fish intake during pregnancy, blood mercury levels, and child cognition at age 3 years in a US cohort. *Am. J. Epidemiol.* 2008;167:1171-81.
[132] Huang LS, Cox C, Myers GJ, et al. Exploring nonlinear association between prenatal methylmercury exposure from fish consumption and child development: evaluation of the Seychelles Child Development Study nine-year data using semiparametric additive models. *Environ. Res.* 2005;97:100-8.
[133] Schantz SL, Gasior DM, Polverejan E, et al. Impairments of memory and learning in older adults exposed to polychlorinated biphenyls via consumption of Great Lakes fish. *Environ. Health Perspect*.2001;109:605-11.
[134] Cott J, Hibbeln JR. Lack of seasonal mood change in Icelanders. *Am. J. Psychiatry* 2001;158:328.
[135] Alessandri JM, Guesnet P, Vancassel S, et al. Polyunsaturated fatty acids in the central nervous system: evolution of concepts and nutritional implications throughout life. *Reprod. Nutr. Dev.* 2004;44:509-38.
[136] Petot GJ, Friedland RP. Lipids, diet and Alzheimer disease: an extended summary. *J. Neurol. Sci.* 2004;226:31-3.
[137] Kalmijn S, van Boxtel MP, Ocke M, Verschuren WM, Kromhout D, Launer LJ. Dietary intake of fatty acids and fish in relation to cognitive performance at middle age. *Neurology* 2004;62:275-80.
[138] Nurk E, Drevon CA, Refsum H, et al. Cognitive performance among the elderly and dietary fish intake: the Hordaland Health Study. *Am. J. Clin. Nutr.* 2007;86:1470-8.
[139] Morris MC, Evans DA, Tangney CC, Bienias JL, Wilson RS. Fish consumption and cognitive decline with age in a large community study. *Arch. Neurol.* 2005;62:1849-53.
[140] van Gelder BM, Tijhuis M, Kalmijn S, Kromhout D. Fish consumption, n-3 fatty acids, and subsequent 5-y cognitive decline in elderly men: the Zutphen Elderly Study. *Am. J. Clin .Nutr* .2007;85:1142-7.
[141] Morris MC, Evans DA, Bienias JL, et al. Consumption of fish and n-3 fatty acids and risk of incident Alzheimer disease. *Arch. Neurol*.2003;60:940-6.
[142] Schaefer EJ, Bongard V, Beiser AS, et al. Plasma phosphatidylcholine docosahexaenoic acid content and risk of dementia and Alzheimer disease: the Framingham Heart Study. *Arch. Neurol.* 2006;63:1545-50.
[143] Solfrizzi V, D'Introno A, Colacicco AM, et al. Dietary fatty acids intake: possible role in cognitive decline and dementia. *Exp. Gerontol.* 2005;40:257-70.
[144] Barberger-Gateau P, Raffaitin C, Letenneur L, et al. Dietary patterns and risk of dementia: the Three-City cohort study. *Neurology* 2007;69:1921-30.
[145] Huang TL, Zandi PP, Tucker KL, et al. Benefits of fatty fish on dementia risk are stronger for those without APOE epsilon4. *Neurology* 2005;65:1409-14.
[146] Barberger-Gateau P, Letenneur L, Deschamps V, Peres K, Dartigues JF, Renaud S. Fish, meat, and risk of dementia: cohort study. *Bmj* 2002;325:932-3.

[147] Donini LM, De Felice MR, Cannella C. Nutritional status determinants and cognition in the elderly. *Arch. Gerontol .Geriatr.* 2007;44 Suppl 1:143-53.
[148] Grant WB. Fish consumption, cancer, and Alzheimer disease. *Am. J. Clin. Nutr.* 2000;71:599-600.
[149] Appleton KM, Woodside JV, Yarnell JW, et al. Depressed mood and dietary fish intake: direct relationship or indirect relationship as a result of diet and lifestyle? *J. Affect Disord.* 2007;104:217-23.
[150] Tanskanen A, Hibbeln JR, Tuomilehto J, et al. Fish consumption and depressive symptoms in the general population in Finland. *Psychiatr .Serv.* 2001;52:529-31.
[151] Hibbeln JR. Fish consumption and major depression. *Lancet* 1998;351:1213.
[152] Timonen M, Horrobin D, Jokelainen J, Laitinen J, Herva A, Rasanen P. Fish consumption and depression: the Northern Finland 1966 birth cohort study. *J. Affect. Disord.* 2004;82:447-52.
[153] Hibbeln JR. Seafood consumption, the DHA content of mothers' milk and prevalence rates of postpartum depression: a cross-national, ecological analysis. *J. Affect. Disord.* 2002;69:15-29.
[154] Browne JC, Scott KM, Silvers KM. Fish consumption in pregnancy and omega-3 status after birth are not associated with postnatal depression. *J. Affect. Disord* .2006;90:131-9.
[155] Miyake Y, Sasaki S, Yokoyama T, et al. Risk of postpartum depression in relation to dietary fish and fat intake in Japan: the Osaka Maternal and Child Health Study. *Psychol .Med.* 2006;36:1727-35.
[156] Domingo JL, Bocio A, Marti-Cid R, Llobet JM. Benefits and risks of fish consumption Part II. RIBEPEIX, a computer program to optimize the balance between the intake of omega-3 fatty acids and chemical contaminants. *Toxicology* 2007;230:227-33.
[157] Herreros MA, Inigo-Nunez S, Sanchez-Perez E, Encinas T, Gonzalez-Bulnes A. Contribution of fish consumption to heavy metals exposure in women of childbearing age from a Mediterranean country (Spain). *Food Chem. Toxicol* .2008;46:1591-5.
[158] Judd N, Griffith WC, Faustman EM. Contribution of PCB exposure from fish consumption to total dioxin-like dietary exposure. *Regul. Toxicol. Pharmacol.* 2004;40:125-35.
[159] Hood E. A diet rich in fish. High-end consumers face more mercury risks. Environ *Health Perspect.* 2003;111:A233.
[160] Storelli MM, Giacominelli-Stuffler R, Storelli A, D'Addabbo R, Palermo C, Marcotrigiano GO. Survey of total mercury and methylmercury levels in edible fish from the Adriatic Sea. *Food Addit. Contam.* 2003;20:1114-9.
[161] Stern AH. Balancing the risks and benefits of fish consumption. *Ann. Intern. Med.* 2005;142:949.
[162] Jeejeebhoy KN. Benefits and risks of a fish diet--should we be eating more or less? *Nat. Clin. Pract. Gastroenterol. Hepatol* .2008;5:178-9.
[163] Kris-Etherton PM, Harris WS, Appel LJ. Fish consumption, fish oil, omega-3 fatty acids, and cardiovascular disease. *Circulation* 2002;106:2747-57.
[164] Arisawa K, Matsumura T, Tohyama C, et al. Fish intake, plasma omega-3 polyunsaturated fatty acids, and polychlorinated dibenzo-p-dioxins/polychlorinated

dibenzo-furans and co-planar polychlorinated biphenyls in the blood of the Japanese population. *Int .Arch. Occup. Environ. Health* 2003;76:205-15.

[165] Gochfeld M, Burger J. Good fish/bad fish: a composite benefit-risk by dose curve. *Neurotoxicology* 2005;26:511-20.

[166] Stern AH. Public health guidance on cardiovascular benefits and risks related to fish consumption. *Environ. Health* 2007;6:31.

[167] Center for Food Safety and Applied Nutrition, US Food and Drug Administration. Consumer advisory: an important message for pregnant women and women of childbearing age who may become pregnant about the risks of mercury in fish. March 2001. Available at: http://www.cfsan.fda.gov/ ~dms/admehg.html. Accessed July 14, 2008.

[168] Domingo JL, Bocio A, Falco G, Llobet JM. Benefits and risks of fish consumption Part I. A quantitative analysis of the intake of omega-3 fatty acids and chemical contaminants. *Toxicology* 2007;230:219-26.

In: Fish Consumption and Health
Editors: George P. Gagne and Richard H. Medrano
ISBN 978-1-60741-151-2

Chapter II

How to Obtain the Health Benefits Associated with Eating Fish from other Sources

William B. Grant [1]
Sunlight, Nutrition, and Health Research Center (SUNARC)
P.O. Box 641603, San Francisco, CA 94164-1603, USA

Abstract

Eating fish, especially cold-water ocean fish, has traditionally been associated with better health, due to generally high amounts of omega-3 fatty acids, vitamin D, and high-quality amino acids. However, due to reduced availability of such fish and an ever increasing world population, many people cannot obtain such fish. Thus, this chapter explores the possibility that the health benefits of eating fish could be obtained from largely vegetable sources. There are good vegetable sources of omega-3 fatty acids, including rapeseed (canola) oil, and flaxseed oil. Vitamin D3 (cholecalciferol) supplements are a reliable source of vitamin D, and both solar and artificial UVB are also good sources in moderation. Vegetable products can provide adequate amounts of protein and essential amino acids, with combining several types of vegetable products indicated for obtaining the essential amino acids in the ratio in which the body can use them.

Introduction

The health benefits of fish consumption are related to the omega-3 fatty acids docosahexaenoic acid (DHA) and eicosapentaenoic acid (EPA), vitamin D, and protein. DHA

[1] www.sunarc.org, wbgrant@infionline.net, 1-415-409-1980.

is an important component of the brain (Dangour and Uauy, 2008), and it has been hypothesized that fish consumption led to increased brain size in humans (Crawford and Harbige, 1988; Crawford et al., 1999; Broadhurst et al., 2002). EPA reduces inflammation (Simopoulos, 2008; Singer et al., 2008), thus countering one of the adverse effects of oxidative stress. Long-chain n-3 polyunsaturated fatty acids (PUFAs) from fish act both directly (e.g., by replacing arachidonic acid as an eicosanoid substrate and inhibiting arachidonic acid metabolism) and indirectly (e.g., by altering the expression of inflammatory genes through effects on transcription factor activation) (Calder, 2006). Vitamin D has many health benefits (Holick, 2007; Grant et al., 2009), and cold water ocean fish (Lu et al., 2007), can be an important source of vitamin D, especially for those living at higher latitudes where solar ultraviolet-B (UVB) doses in winter are insufficient for vitamin D production (Jablonski and Chaplin, 2000). Fish protein is generally good but is somewhat deficient in essential amino acids such as phenylalanine.

As good as the health benefits of fish are, there are a number of reasons why not everyone can or wants to eat fish. First, high-quality cold-water ocean fish, which have the best fatty acid profiles (Philibert, Vanier, Abdelouahab, Chan and Mergler, 2006) and higher vitamin D levels (Lu et al., 2007) are relatively expensive. Second, fish stocks are being rapidly depleted. It is estimated that the large predatory fish biomass today is only about 10% of pre-industrial levels (Myers and Worm, 2003, 2005). This trend has adverse effects down the food chain. For example, as sharks are killed in high numbers, their prey fish can increase, which, in the case of the cownose ray, led to the termination of a century-long bay scallop fishery (Myers, Baum, Shepherd, Powers, and Peterson, 2007). Third, bringing cold-water ocean fish to market generally requires large amounts of fossil fuel for transport and storage at cold temperatures. Fourth, the larger cold-water ocean fish and fresh-water fish in polluted waters tend to have higher levels of mercury (Miller et al., 1972) and organic pollutants (Hylland et al., 2006). Fifth, many people prefer to eat plant-based foods rather than animal-based foods (Lea and Worsley, 2002). Sixth, plant-based foods are generally a more energy-efficient source of food than animal-based foods since plant matter has to be converted to animal matter. For example, fish require about 0.36 g of protein for each g of protein of fish produced (Swann, 2004).

The adverse health effects of mercury are well known. Mercury primarily affects the neurological system including the brain, especially during fetal development. A recent study of mercury in cord blood at x months found a reduction of 15 IQ points for a mean cord mercury level of 7.74 μg/L compared to 0.1 μg/L (Lederman et al., 2008). However, 7.74 μg/L is a very high number that few would experience.

A recent summary of the health effects of eating fish reached this conclusion (Mozaffarian and Rimm, 2006):

> "Modest consumption of fish (eg, 1-2 servings/wk), especially species higher in the n-3 fatty acids eicosapentaenoic acid (EPA) and docosahexaenoic acid (DHA), reduces risk of coronary death by 36% (95% confidence interval, 20%-50%; P<.001) and total mortality by 17% (95% confidence interval, 0%-32%; P = .046) and may favorably affect other clinical outcomes. Intake of 250 mg/d of EPA and DHA appears sufficient for primary prevention. DHA appears beneficial for, and low-level methylmercury may adversely affect, early neurodevelopment. Women of childbearing age and nursing

mothers should consume 2 seafood servings/wk, limiting intake of selected species. Health effects of low-level methylmercury in adults are not clearly established; methylmercury may modestly decrease the cardiovascular benefits of fish intake. A variety of seafood should be consumed; individuals with very high consumption (> or =5 servings/wk) should limit intake of species highest in mercury levels. Levels of dioxins and polychlorinated biphenyls in fish are low, and potential carcinogenic and other effects are outweighed by potential benefits of fish intake and should have little impact on choices or consumption of seafood (women of childbearing age should consult regional advisories for locally caught freshwater fish)."

However, as pointed out by Jane M. Hightower in her new book *Diognasis Mercury* (2009), the message this review gave was that adults did not have to worry about eating fish. Unfortunately, as she has learned from her practice in San Francisco, many high-end people eat so much top predator ocean fish that they develop significant adverse health effects.

A recent study of blood mercury (BHg) levels among women of childbearing age in the United States, with a level of BHg>3.5 μg/L considered of concern, found the following:

"Elevated BHg occurred more commonly among women of childbearing age living in coastal areas of the United States (approximately one in six women). Regionally, exposures differ across the United States: Northeast > South and West > Midwest. Asian women and women with higher income ate more fish and had higher BHg."

An alternative to wild fish is raising fish by aquaculture. However, aquaculture has its share of problems too. For one, wild salmon and sea trout populations are declining due to recurrent louse infestations from farm salmon (Krkosek et al., 2007; Ford and Myers, 2008). Second, farmed salmon can have higher organic pollutant concentrations (Foran et al., 2005; Hamilton et al., 2005; Shaw et al., 2006) and lower omega-3 fatty acid levels than wild salmon (Ikonomou et al., 2007; Friesen, Ikonomou, Higgs, Ang, and Dubetz, 2008). Third, in many cases, large quantities of smaller fish are harvested as feed for the farmed fish. Fourth, farmed fish may be fed vegetable seeds or grains, which may not provide a good fatty acid profile for the fish. Nonetheless, given the rapidly deteriorating state of our ocean's fish resources, aquaculture will continue to grow in importance, and the problems encountered will be overcome through better practices.

As alternatives to fish for the health benefits of fish, this chapter explores vegetable sources of omega-3 fatty acids and protein and solar and artificial UVB and supplements for vitamin D.

Health Benefits of Fish

Omega-3 Fatty Acids

Since authors of the other chapters of this book will likely discuss the health benefits of omega-3 fatty acids for which they are most familiar, I will do likewise, discussing Alzheimer's disease and cancer.

Alzheimer's Disease

My first paper on determining the risk factors for chronic disease was on dietary links for Alzheimer's disease (Grant, 1997). I took an interest in the topic in 1996 upon learning that Japanese-American men living in Hawaii had 2.5 times the prevalence of Alzheimer's disease compared to native Japanese (White et al., 1996). During the previous three years I had been studying the effect of acid rain and ozone on Eastern U.S. hardwood forests as a volunteer project for the Sierra Club (later, Appalachian Voices) when working for the NASA Langley Research Center in Hampton, VA, and acquired a familiarity with the ecologic approach under the tutelage of Professor Orie Loucks (Miami University, Oxford, Ohio). I also learned that in forests affected by acid rain, as the pH is lowered (greater acidity), the base cations (positively charged ions) such as calcium, magnesium and potassium are leached from the soil, while aluminum and transition metal (copper, iron, mercury, manganese, zinc, etc.) oxides are dissolved and these ions enter the soil solution where they can be easily taken up by trees and other plants. Thus, the effect of acid rain on trees is to reduce the availability of base cations and replace them with aluminum and transition metal ions. As a result of the trace metal ion replacements, trees go into decline and early senescence. This work led to the production of the Sierra Club book, *An Appalachian Tragedy* (Ayers, Hager, and Little, 1998).

I was also aware that those with Alzheimer's disease had elevated concentrations of aluminum in their brains. I then made the leap to the hypothesis that since the primary environmental effect for most people is what they eat, that if I could do an ecological study of Alzheimer's disease prevalence by country and look at correlations with major components of the national diets, I could determine the major risk factors for Alzheimer's disease. I found appropriate data for 11 countries and did the analysis. The primary risk factors turned out to be total energy (caloric) and fat supply, while the most important risk reduction factor was cereals or grains. In addition, for the set of countries with higher amounts of fat in the national diet, fish reduced the risk (Grant, 1997). I also learned that studies of the brains of those who died with Alzheimer's disease had below average base cation concentrations and above average transition metal ion concentrations (Ehmann, Markesbery, Alauddin, Hossain, and Brubaker, 1986). My hypothesis was that a diet tending to produce a lower pH in the digestive tract had an action similar to that in the forest soils. Ongoing work has supported the hypothesis that transition metal ions are an important risk factor for Alzheimer's disease through the mechanism of free radical generation and oxidative stress (Smith, 1995; Smith, Rottkamp, Nunomura, Raina and Perry, 2000; Grant, 1997; Markesbery, 1997). While my paper elicited the questioning upon publication generally afforded a novel hypothesis by those entrenched in an old field, I continued writing on the topic (Grant, 1998, 1999, 2000; Grant, Campbell, Itzhaki, and Savory, 2002). Once several observational studies (Barberger-Gateau et al., 2002; Luchsinger, Tang, Shea, and Mayeux, 2002; Morris et al., 2003) supported the diet/Alzheimer's disease hypothesis, it was generally accepted. The beneficial role of fish in reducing the risk of Alzheimer's disease is now well supported in observational studies (Morris, Evans, Tangney, Bienias, and Wilson, 2005; Huang et al., 2005; Schaefer et al., 2006; Barberger-Gateau et al., 2007) as well in the understanding of the mechanisms (Ma et al., 2007; Farooqui, Horrocks, and Farooqui, 2007; Farooqui, Ong, Horrocks, Chen, and Farooqui, 2007).

Cancer

I have also examined the role of fish consumption in reducing the risk of cancer. In a multi-country study of breast cancer, I found that fish supply in the national diet was inversely correlated with breast cancer mortality rates in a multiple linear regression analysis in which the ratio of animal products to vegetable products and alcohol were directly correlated (Grant, 2002). A summary of case-control studies conducted in northern Italy between 1983 and 1996 reported the following:

> "There was a consistent pattern of protection against the risk of digestive tract cancers with fish consumption: oral cavity and pharynx, OR = 0.5 for the highest compared with the lowest level of consumption; esophagus, OR = 0.6; stomach, OR = 0.7; colon, OR = 0.6; rectum, OR = 0.5; and pancreas, OR = 0.7. There were inverse trends in risk of larynx (OR = 0.7), endometrial (OR = 0.8), and ovarian (OR = 0.7) cancers and multiple myeloma (OR = 0.5). No pattern of cancer risk in relation to fish consumption was observed for cancers of the liver, gallbladder, breast, bladder, kidney, or thyroid or for lymphomas." (Fernandez, Chatenoud, La Vecchia, Negri, and Franceschi, 1999).

Whether the beneficial effects were due to omega-3 fatty acids or vitamin D was not clear, with both likely playing a role (Grant, 2000).

In a cursory search of papers reporting correlations between fish consumption and cancer risk, a number of papers were found. Fish consumption has also been reported protective against breast (Gago-Dominguez, Yuan, Sun, Lee, and Yu, 2003; Wakai et al., 2005), colorectal (Yang et al., 2003; Norat et al., 2005; Geelen et al., 2007; Hall, Chavarro, Lee, Willett, and Ma, 2008), endometrial (Terry, Wolk, Vaini, and Weiderpass, 2002), gastric (Hall et al., 2008), lung (Takezaki et al., 2003), nasopharyngeal (Feng et al., 2007), prostate (Augustsson et al., 2003; Pham et al., 2008), renal (Wolk, Larsson, Johansson, and Ekman, 2006) cancer, melanoma (Fortes et al., 2008), leukaemia (Fritschi, Ambrosini, Kliewer, and Johnson, 2004), multiple myeloma (Fritschi et al., 2004; Hosgood, Baris, Zahm, Zheng, and Cross, 2007), and non-Hodgkin lymphoma (Fritschi et al., 2004). No effect was found for bladder (Holick, Giovannucci, Stampfer, and Michaud, 2006), breast (Engeset et al., 2006), pancreatic (Larsson, Håkanson, Permert, and Wolk, 2006), and thyroid (Bosetti et al., 2001) cancer. A positive correlation was found for lung cancer (Dosil-Díaz, Ruano-Ruiz, Gestal-Otero, and Barros-Dios, 2007). I have also suggested that American Indians living in Alaska have a higher rate of cancer than those living in New Mexico because their diet now has much less fish than their traditional diet (Grant, 2007). Thus, there is reasonable evidence that eating fish reduces the risk of many types of cancer.

The fact that fish consumption has been found to have little effect for several cancers or an adverse effect for lung cancer suggests that fish have several components which may increase the risk of cancer. For example, salt-preserved fish is a risk factor for colorectal (Yang et al., 2003), gastric (Strumylaite, Zickute, Dudzevicius, and Dregval, 2006) and lung (Takezaki et al., 2003) cancer.

Alternative Sources for the Health Benefits of Eating Fish

Vegetables as Sources of Omega-3 Fatty Acids

Fish obtain their omega-3 fatty acids from algae or by eating other fish. Algae have been exploited as a source of omega-3 fatty acids for humans (Doughman, Krupanidhi, and Sanjeevi, 2007; Arterburn et al., 2008). However, they have not reached the stage of being a large sustainable source of omega-3 fatty acids.

Other common sources of omega-3 fatty acids, primarily alpha-linolenic acid (18:3ω3, ALA), are flax (lin) seeds and oils (Rao, Abdel-Reheem, Bhella, McCracken, and Hildebrand, 2008), rapeseed (canola) (10% ALA) oil, soybean (6.5% ALA), and mustard (9.5% ALA), shia and walnuts (Bourre, 2007). Flaxseed (55% ALA in the oil) has been found to confer modest reduction of cardiovascular risk factors (Bloedon et al., 2008). It was found that when Eastern European countries changed a large portion of their fat intake from animal fat to rapeseed oil, coronary heart disease rates decreased dramatically (Zatonski, Campos, and Willett, 2008). This effect was not found from zero% ALA sunflower seed oil. Substituting walnuts for monounsaturated fat in a Mediterranean diet improves endothelium-dependent vasodilation in hypercholesterolemic subjects (Ros et al., 2004). There are several additional good references on omega-3 fatty acids and cardiovascular risk factors (von Schacky, 2997; Kaul, 2008; Lee, 2008).

A still esoteric vegetable source of omega-3 fatty acid ALA is common purslane [*Portulaca oleracera*] (Simopoulos and Salem, 1986; Simopoulos, Norman, Gillaspy, and Duke, 1992). Purslane contains 3-4 mg of omega-3 fatty acids per gram of wet weight (Simopoulos and Salem, 1986; Simopoulos et al., 1992). The next best source, spinach, has 0.5-0.9 mg per gram (Simopoulos and Salem, 1986). Purslane generally grows wild in Europe and the United States. It can be used in salads raw or cooked as a vegetable. A 100-g serving of purslane contains 300-400 mg of 18:3ω3, 12 mg of α-tocopherol, 27 mg of ascorbic acid, 2 mg of β-carotene, and 15 mg of glutathione, making it a good dietary source of omega-3 fatty acids and antioxidants (Simopoulos et al., 1992; Liu et al., 2000).

For those who want an animal source of omega-3 fatty acids other than fish, eggs from chickens that consumed plant sources of omega-3 fatty acids is a possibility. It was found that eggs from free-range chickens in Greece had higher ALA than most eggs. The source of the ALA was traced to foraging on purslane (Simopoulos and Salem, 1986; Simopoulos et al., 1992; Simopoulos, 2004a).

As mentioned, DHA is very important for proper brain development (Dangour and Uauy, 2008). Since the majority of people do not have access to fish with DHA, the question arises how pregnant and lactating women are able to supply the developing fetus and nursing infant with DHA. It appears that women are able to convert ALA to DHA. A fractional conversion of 9% was measured (Williams and Burdge, 2006). The better conversion by women than men may be due to the role played by estrogen (Burdge and Wootton, 2002; Burdge and Calder, 2005; Williams and Burdge, 2006).

It is also important to try to achieve a high dietary omega-3 to omega-6 fatty acid ratio, i.e., more omega-3, less omega-6 (Simopoulos, 1999a, 1999b, 2004b, 2008). Omega-6 fatty acids, largely represented by linolenic acid (18:2ω6; LA), are mainly found in vegetable oils made from seeds such as sunflower seeds (70% LA), safflower (75% LA), soy (50% LA), walnut (57% LA), peanut (35% LA), cottonseed (54%LA), grapeseed (70% LA), corn (58% LA, and sesame seeds (45% LA). The adverse effects of omega-6 fatty acids include increasing the oxidation of low-density lipoprotein and severity of coronary atherosclerosis, inhibition of incorporation of EPA, increasing platelet aggregation, increasing biomarkers of inflammation, and possibly promoting cancer. A recent paper reported that high dietary omega-6 fatty acids contribute to reduced DHA in the developing brain and inhibit secondary neurite growth (Novak, Dyer, and Innis, 2008). It is estimated that in the paleo diet, the ratio of omega-3 to omega-6 fatty acids was 1:0.79 (Eaton, Eaton, Sinclair, Cordain, and Mann, 1998); currently in Western diets, the ratio is estimated to be from 8 to 20.

Vitamin D

The health benefits of vitamin D are receiving much attention lately. In many northern societies, marine fish and especially fish liver oil was often the only available source (Jablonski and Chaplin, 2000). In the past two decades, many non-calcemic benefits such as cancer prevention have been identified and studied using a variety of approaches including ecological, observational, interventional, and laboratory studies. A summary of the history of discoveries of beneficial roles and recent reports is presented in Table 1. An effort was made to identify the first paper reporting a link between solar UVB irradiance or vitamin D and reduced risk of disease. Some early papers may have been overlooked. As can be seen, many of the beneficial effects of vitamin D have been identified only very recently.

Many of the studies were ecological or observational, with only a few randomized controlled studies (RCTs). Since man has always lived in harmony with the sun and solar UVB, it is reasonable that well-designed ecological and observational studies can determine the vitamin D/health links and quantify them to some extent. RCTs are expensive and time consuming and not always properly designed. Many of the early colorectal cancer RCTs used too little vitamin D (Grant and Garland, 2006; Gorham et al., 2007; Garland et al., 2007). Even the Women's Health Initiative study of vitamin D and calcium supplements and risk of colon cancer and fractures used only 400 International Units (IU) of vitamin D and found very few effects (Jackson et al., 2006; Wactawski-Wende et al., 2006; Ding, Mehta, Fawzi, and Giovannucci, 2007). However, when larger amounts are used, strong beneficial effects have been found for fractures (Bischoff-Ferrari et al., 2005, 2007), influenza and the common cold (Aloia and Li-Ng, 2007) and cancer (Lappe, Travers-Gutafson, Davies, Recker, and Heaney, 2007). A meta-analysis of RCTs found that a mean intake of 528 IU/day for a duration of 5.7 years was associated with a relative risk for mortality from any cause of 0.93 (95% confidence interval, 0.87-0.99). (Autier and Gandidi, 2007). A recent study in the United States found in multivariate models (adjusted for baseline demographics, season, and traditional and novel CVD risk factors), compared with the highest quartile, being in the lowest quartile [25(OH)D levels <17.8 ng/mL; times 2.5 gives the common nmol/L scale, i.e., <44.5 nmol/L in this case] was associated with a 26% increased rate of all-cause mortality (mortality rate ratio, 1.26; 95% CI, 1.08-1.46) (Melamed, Michos, Post, and Astor, 2008).

Good reviews of the health benefits of vitamin D are found in several recent papers (Grant and Holick, 2005; Holick, 2007; Schwalfenberg, 2007; Anonymous, 2008; Cannell and Hollis, 2008; Cannell, Hollis, Zasloff, and Heaney, 2008; Cherniack, Florez, Roos, Troen, and Levis, 2008). Based on many studies, the recommended daily production or intake of vitamin D is at least 1000-2000 IU (Grant and Holick, 2005; Grant and Garland 2006; Aloia and Li-Ng, 2007; Lappe et al., 2007) up to 3600 IU (Heaney, Davies, Chen, Holick, and Barger-Lux, 2003) based on a desirable serum 25(OH)D level of 30-60 ng/mL (Grant and Holick, 2005; Holick, 2007). In addition, pregnant and nursing women require up to 6000 IU/day (Hollis, 2007). It would be difficult to obtain these amounts of vitamin D from consuming seafood unless seafood was the main component of the diet, such as for the Inuit (Kuhnlein and Receveur, 2007; Kuhnlein, Receveur, Soueida, and Berti, 2008).

Vitamin D can be obtained from three basic sources: natural or artificial UVB irradiance, natural or fortified foods, and supplements. Solar UVB is the primary source of vitamin D for most people. However, when the solar elevation angle is lower than about 45°, i.e., when your shadow is longer than your height, there is too little UVB to efficiently generate vitamin D. A study in Boston found that there are 4-5 months a year when it is impossible to produce vitamin D (Webb, Kline, and Holick, 1988). A recent publication calculated the amount of vitamin D that can be produced as a function of latitude and day of the year (Webb, 2006; Webb and Engelsen, 2006). It should be noted that vitamin D production efficiency decreases with age, and a person over the age of 60 years makes vitamin D at 25% the rate of one under the age of 20 years (MacLaughlin and Holick, 1985). Artificial UVB sources can be used to produce about 10,000 IU in a few minutes (Porojnicu, Bruland, Aksnes, Grant, and Moan, 2008).

Food including fish and fortified foods such as milk and orange juice supply the average American with 250-300 IU of vitamin D3/day (McCullough et al., 2003). This is a very insignificant amount based on many recent studies. However, it is better than nothing, which is the general rule in most European countries. More food should be fortified with vitamin D (Calvo, Whiting, and Barton, 2004; Calvo and Whiting, 2006), and cereal/grain products such as bread are good foods to fortify (Natri et al., 2006).

Vitamin D supplements represent the most efficient and often least expensive way to obtain vitamin D. It appears that the body can use about 3600 IU of vitamin D per day (Heaney et al., 2003), which is more than most people can produce from solar UVB with today's lifestyles; it would take nearly half of the body exposed for times from a few minutes to an hour or more, depending on season, latitude, time of day, age, and skin pigmentation. Most vitamin D is available as cholecalciferol (vitamin D3) produced from UVB-irradiated sheep's lanolin, while some is obtained from fish oil or cod liver oil. Vitamin D is also available as ergocalciferol (vitamin D2), made from plant based sources such as yeast and mushrooms. While vitamin D2 raises serum 25(OH)D levels as well as vitamin D3 does (Holick et al., 2008), there is a concern that vitamin D2 is less effective in improving health (Houghton and Vieth, 2006). When selecting supplements, it is advised to limit vitamin A intake to 1500 IU/day, as vitamin A competes with vitamin D, thereby reducing calcium intake and increasing hip fracture risk (Melhus et al., 1998; Feskanich, Singh, Willett, and Colditz, 2002).

In general, the adverse effects of vitamin D start well above 10,000 IU/day (Hathcock, Shao, Vieth, and Heaney, 2007). However, those with various diseases may suffer from hypercalcaemia and calcification of the arteries for too much vitamin D, including those with granulomatous diseases such as sarcoidosis (Hewison et al., 2007), and some with lymphoma (Hewison et al., 2003). People with such diseases should carefully monitor their 1,25-dihydroxyvitamin D levels if they supplement with high doses.

Amino Acids in Fish

Our protein is comprised of 20 amino acids (AAs). There are about eight AAs that are called "essential amino acids" (EAA) because the human body cannot synthesize them from other AAs. The EAAs are generally utilized by the body in the same ratio as found in egg protein. Thus, if some of EAAs are in short supply, they will limit the utilization of the other AAs. Frances Moore Lappé realized this and proposed the concept of protein complementarity in her 1971 book, *Diet for a Small Planet*. Her thesis was that the earth could not support a large population if everyone obtained their protein from animal products, but could if they combined plant sources of protein in a manner so as to obtain all of the essential amino acids in approximation to the ratios in egg protein. While she suggested that the proteins should be combined in the same meal, she relaxed that requirement later when she realized that the body stores AAs for some time and can combine them from different meals.

A good source of nutrient content of foods is the USDA National Nutrient Database for Standard Reference (United States Department of Agriculture, 2007). Data for representative dietary protein sources can be obtained. The data for EAAs are given in Table 2. Fish is low in phenylalanine but high in lysine while rice is low lysine, so combining fish and rice yields a good AA balance. Many vegetable sources are deficient in lysine, although soybeans have adequate amounts. Vegetarians are more likely than non-vegetarians to have hyperhomocysteinemia, with low vitamin B12 intake contributing as a deleterious factor (Krajcovicova-Kudlackova et al., 2007).

While the total amino acid content per 100 g of vegetable products is much lower than that of animal products, the difference need not be important. National diets generally derive about 10-13% of their energy from protein; those with the higher protein fraction generally have higher amounts of animal products and better health. In 1992-4, protein fractions of total dietary supply were: Argentina: 0.126; Australia: 0.132; Brazil: 0.097; Germany: 0.116; Hungary: 0.105; India: 0.097; Japan: 0.131; Pakistan: 0.098; United Kingdom: 0.113; United States: 0.125 (Food and Agriculture Organization, 1996). However, in the period 1961-3, Japan obtained 11.7% of its energy from protein and 10.4% from animal products, and in the 1960s and 1970s, chronic disease rates were among the lowest in the world and healthy life expectancy was at or near the top. In the period 1992-4, 31% of food energy was derived from animal products, and chronic disease rates increased (Food and Agriculture Organization., 1996). Thus, a largely vegetarian or vegan diet can supply all of the required protein and be a very healthy diet. The limitations of a largely vegetarian diet generally include low protein, low omega-3 fatty acids (Kornsteiner, Singer, and Elmadfa, 2008), and low vitamin B12 (Stabler and Allen, 2004).

However, rice is relatively protein poor for cereals and grains with about 8% protein per 100 g of rice. Low dietary protein is a health risk for a number of diseases. Countries with more than 50% of their energy derived from cereals such as rice obtain about 10% of their energy from protein. Such countries have poorer health as a result. Countries such as India and Pakistan have high rates of diabetes (Ramachandran, 2005) and coronary heart disease and risk factors in India (Gupta et al., 2002, 2004), and lower life expectancies (53.5 years compared to 69 years for the United States) (World Health Organization, 2008).

A study in Japan found white rice (GI = 77) was the major contributor to dietary glycemic index (GI) and glycemic load (GL) (58.5%). A meta-analysis of observational studies found for the comparison between the highest and lowest quantiles of GI and GL, significant positive associations were found in fully adjusted models of validated studies for type 2 diabetes (GI RR = 1.40, 95% CI: 1.23, 1.59; GL RR = 1.27, 95% CI: 1.12, 1.45), coronary heart disease (GI RR = 1.25, 95% CI: 1.00, 1.56), gallbladder disease (GI RR = 1.26, 95% CI: 1.13, 1.40; GL RR = 1.41, 95% CI: 1.25, 1.60), breast cancer (GI RR = 1.08, 95% CI: 1.02, 1.16), and all diseases combined (GI RR = 1.14, 95% CI: 1.09, 1.19; GL RR = 1.09, 95% CI: 1.04, 1.15) (Barclay et al., 2008).

Increased protein consumption while growing increases adult height. Data from rural provinces of China from the early 1980s provide some evidence of this idea. Those in the northern provinces eat wheat as the staple grain, while those in the southern provinces eat rice as the staple grain. Those in the northern provinces had a mean height of about 160 cm, while those in the southern provinces had a mean height of 157 cm (Chen, Campbell, Li, and Peto, 1990).

Summary and Conclusion

Eating fish, especially cold-water ocean fish, has traditionally been associated with better health, due to generally high amounts of omega-3 fatty acids, vitamin D, and good amino acids. However, due to reduced availability of such fish and an ever-increasing world population, many people cannot obtain such fish. Thus, this chapter explored the possibility that the health benefits of eating cold-water ocean fish could be obtained from largely vegetable sources. There are good vegetable sources of omega-3 fatty acids, including canola (rapeseed) oil and flaxseed oil. Vitamin D supplements, largely obtained from UVB-irradiated sheep's lanolin, are a reliable source of vitamin D, and both solar and artificial UVB are also good sources if used properly. Vegetable products can provide adequate amounts of protein and essential amino acids, with combining several types of vegetable products indicated for obtaining the essential amino acids in the ratio in which the body can use them.

Disclosure

I receive funding from the UV Foundation (McLean, VA), the Vitamin D Society (Canada), and the European Sunlight Association (Brussels).

Table 1. Chronology of identification of vitamin D–sensitive diseases from representative early mention to recent strong finding or review*

Disease	Early mention	Recent report
Rickets	Sniadecki and Sniadecki, 1840	Martins e Silva, 2007
Tuberculosis	Rollier, 1927	Liu, Stenger, Tang, and Modlin, 2007
Dental caries	East, 1939	
Osteopenia, osteoporosis and osteomalacia	De Seze, Lichtwitz, Hioco, and Pordier, 1955	Holick, 2006
Preeclampsia	Glatzel, 1969	Bodnar et al., 2007
Cardiovascular disease	Lund, Badskjaer, Lund, and Soerensen, 1978	Dobnig et al., 2008; Giovannucci et al., 2008; Pilz et al., 2008a; Wang et al., 2008
Cancer	Garland and Garland, 1980	Garland et al., 2006; Grant, 2002; Giovannucci et al., 2006; Grant and Garland, 2006; Lappe et al., 2007; Grant, 2009a
Seasonal influenza	Hope-Simpson, 1981	Cannell et al., 2006; Aloia and Li-Ng, 2007
Congestive heart failure	McGonigle, Fowler, Timmis, Weston, and Parsons, 1984	Zittermann et al., 2008; Pilz et al., 2008b
Type 2 diabetes mellitus	Ishida et al., 1985	Pittas et al., 2006; Pittas, Lau, Hu, and Dawson-Hughes, 2007; Knekt et al., 2008
Multiple sclerosis	Goldberg, Fleming, and Picard, 1986	Munger, Levin, Hollis, Howard, and Ascherio, 2006
Hypertension	Rostand, 1997	Forman et al., 2007
Pneumonia	Muhe, Lulseged, Mason, and Simoes, 1997	Roth, Jones, Prosser, Robinson, and Vohra, 2008
Chronic obstructive pulmonary disease	Schellenberg et al., 1998	Black and Scragg, 2005
Schizophrenia, risk in utero	McGrath, 1999	Edwards, 2007
Type 1 diabetes mellitus	Anonymous, 1999	Zipitis and Akobeng, 2008
Periodontal disease	Dietrich, Joshipura, Dawson-Hughes, and Bischoff-Ferrari, 2004	Park, Nam, and Choi, 2006
Asthma	Poon et al., 2004	Ginde AA, Mansbach JM, and Camargo, 2009
Septicemia	Fukumoto et al., 2005	Mookherjee, Rehaume, and Hancock, 2007; Grant, 2009b
Depression	Wilkins, Sheline, Roe, Birge, and Morris, 2006	Hoogendijk et al., 2008
Dementia		Buell and Dawson-Hughes, 2008; Grant, 2009c

* Adapted from Grant (2009d)

Table 2. Essential amino acid content, g/100 g edible portion [USDA, 2007). Numbers in () indicate limiting EAAs, defined here as <75% of the mean of the ratios of each EAA to that of eggs, while * indicates >125% of the mean

Amino acid	Egg whites	Salmon	Nonfat milk	English walnuts	Sesame seeds	Peanuts	Soybeans	Rice brown	Wheat drum
Tryptophan	1.00	0.26	0.51	0.17*	0.39*	0.23	0.51*	0.10	0.18*
Threonine	3.69	1.03	1.63	0.60	0.74	0.81	1.53	0.29	0.37
Isoleucine	4.58	1.08	2.19	0.63	0.76	0.83	1.71	0.34	0.53
Leucine	6.84	1.91	3.54	1.17*	1.36	1.54	2.87	0.66	0.93
Lysine	5.52	2.15*	2.87	(0.42)	(0.57)	(0.85)	2.34	(0.30)	(0.30)
Methionine	2.79	0.69	(0.91)	(0.24)	0.59	(0.29)	(0.48)	(0.18)	(0.22)
Phenylalanine	4.74	(0.92)	(1.75)	0.71	0.94	1.23*	1.84	0.41	0.68*
Valine	5.16	1.21	2.42	(0.41)	0.99	0.99	1.76	0.47	0.59
Total protein	81.1	23.5	36.2	15.2	17.7	23.7	35.2	7.9	13.7

References

Aloia JF, and Li-Ng M. (2007) Re: epidemic influenza and vitamin D. *Epidemiol. Infect.*, 135:1095-6; author reply 1097-8.

Anonymous. (1999) Vitamin D supplement in early childhood and risk for Type I (insulin-dependent) diabetes mellitus. The EURODIAB Substudy 2 Study Group. *Diabetologia.* 42:51-4.

Anonymous. (2008) Vitamin D - monograph. *Altern. Med. Rev.* 13:153-64.

Arterburn LM, Oken HA, Bailey Hall E, Hamersley J, Kuratko CN, and Hoffman JP. (2008) Algal-oil capsules and cooked salmon: nutritionally equivalent sources of docosahexaenoic acid. *J. Am. Diet .Assoc.* 108:1204-9.

Augustsson K, Michaud DS, Rimm EB, Leitzmann MF, Stampfer MJ, Willett WC, et al. (2003) A prospective study of intake of fish and marine fatty acids and prostate cancer. *Cancer Epidemiol. Biomarkers Prev.* 12:64-7.

Autier P, and Gandini S. (2007) Vitamin D Supplementation and Total Mortality: a Meta-analysis of Randomized Controlled Trials. *Arch. Intern. Med.* 167:1730-7.

Ayers H, Hager J, and Little CE. (1998) *Appalachian Tragedy, Air Pollution and Tree Death in the Eastern Forests of North America.* Sierra Club Books, San Francisco. .

Barberger-Gateau P, Letenneur L, Deschamps V, Peres K, Dartigues JF, and Renaud S. (2002) Fish, meat, and risk of dementia: cohort study. *BMJ;*325:932-3.

Barberger-Gateau P, Raffaitin C, Letenneur L, Berr C, Tzourio C, Dartigues JF, et al. (2007) Dietary patterns and risk of dementia: the Three-City cohort study. *Neurology.* 69:1921-30.

Barclay AW, Petocz P, McMillan-Price J, Flood VM, Prvan T, Mitchell P, et al. (2008) Glycemic index, glycemic load, and chronic disease risk--a meta-analysis of observational studies. *Am. J. Clin. Nutr.* 87:627-37.

Bischoff-Ferrari HA, Dawson-Hughes B, Baron JA, Burckhardt P, Li R, Spiegelman D, et al. (2007) Calcium intake and hip fracture risk in men and women: a meta-analysis of prospective cohort studies and randomized controlled trials. *Am. J. Clin. Nutr.* 86:1780-90.

Bischoff-Ferrari HA, Willett WC, Wong JB, Giovannucci E, Dietrich T, and Dawson-Hughes B. (2005) Fracture prevention with vitamin D supplementation: a meta-analysis of randomized controlled trials. *JAMA*. 293:2257-64.

Black PN, and Scragg R. (2005) Relationship between serum 25-hydroxyvitamin d and pulmonary function in the third national health and nutrition examination survey. *Chest*;128:3792-8.

Bloedon LT, Balikai S, Chittams J, Cunnane SC, Berlin JA, Rader DJ, et al. (2008) Flaxseed and cardiovascular risk factors: results from a double blind, randomized, controlled clinical trial. *J. Am. Coll. Nutr.* 27:65-74.

Bodnar LM, Catov JM, Simhan HN, Holick MF, Powers RW, and Roberts JM. (2007) Maternal vitamin D deficiency increases the risk of preeclampsia. *J. Clin. Endocrinol. Metab.*92:3517-22.

Bosetti C, Kolonel L, Negri E, Ron E, Franceschi S, Dal Maso L, et al. (2001) A pooled analysis of case-control studies of thyroid cancer. VI. Fish and shellfish consumption. *Cancer Causes Control.* 12:375-82.

Bourre JM. (2007) Dietary omega-3 fatty acids for women. *Biomed Pharmacother*. 61:105-12.

Broadhurst CL, Wang Y, Crawford MA, Cunnane SC, Parkington JE, and Schmidt WF. (2002) Brain-specific lipids from marine, lacustrine, or terrestrial food resources: potential impact on early African Homo sapiens. *Comp. Biochem. Physiol. B. Biochem. Mol. Biol.* 131:653-73.

Buell JS, and Dawson-Hughes B. (2008) Vitamin D and neurocognitive dysfunction: Preventing "D"ecline? *Mol. Aspects Med.* 29:415-22.

Burdge GC, and Calder PC. (2005) Conversion of alpha-linolenic acid to longer-chain polyunsaturated fatty acids in human adults. *Reprod. Nutr. Dev.* 45:581-97.

Burdge GC, and Wootton SA. (2002) Conversion of alpha-linolenic acid to eicosapentaenoic, docosapentaenoic and docosahexaenoic acids in young women. *Br. J. Nutr.* 88:411-20.

Calder PC. (2006) n-3 polyunsaturated fatty acids, inflammation, and inflammatory diseases. *Am. J. Clin. Nutr.* 83 Suppl:1505S-1519S.

Calvo MS, Whiting SJ, and Barton CN. (2004) Vitamin D fortification in the United States and Canada: current status and data needs. *Am. J. Clin. Nutr.* 80 Suppl:1710S-6S.

Calvo MS, and Whiting SJ. (2006) Public health strategies to overcome barriers to optimal vitamin D status in populations with special needs. *J. Nutr.* 136:1135-9.

Cannell JJ, Vieth R, Umhau JC, Holick MF, Grant WB, Madronich S, et al. (2006) Epidemic influenza and vitamin D. *Epidemiol. Infect.*134:1129-40.

Cannell J, Hollis B, Zasloff M, and Heaney R. (2008) Diagnosis and treatment of vitamin D deficiency. *Expert Opin. Pharmacother.* 9:107-18.

Cannell JJ, and Hollis BW. (2008) Use of vitamin D in clinical practice. *Altern. Med. Rev.* 13:6-20.

Chen J, Campbell TC, Li J, and Peto R (1990) *Diet, Life-style and Mortality in China*. Oxford University Press

Cherniack EP, Florez H, Roos BA, Troen BR, and Levis S. (2008) Hypovitaminosis D in the elderly: from bone to brain. *J. Nutr. Health Aging.* 12:366-73.

Crawford MA, Bloom M, Broadhurst CL, Schmidt WF, Cunnane SC, Galli C, et al. (1999) Evidence for the unique function of docosahexaenoic acid during the evolution of the modern hominid brain. *Lipids.*34 Suppl:S39-47.

Crawford MA, and Harbige LS. (1988) n-3 fatty acids and the evolution of the brain. *Prog. Clin. Biol. Res.*282:335-54.

Dangour AD, and Uauy R. (2008) N-3 long-chain polyunsaturated fatty acids for optimal function during brain development and ageing. *Asia Pac. J. Clin. Nutr.*17 Suppl 1:185-8.

De Seze S, Lichtwitz A, Hioco D, and Pordier P. (1955) [Modifications of calcium in urine caused by vitamin D in osteomalacia of supply and osteoporosis.] *Bull. Mem. Soc. Med. Hop.* Paris.71:961-6. French.

Dietrich T, Joshipura KJ, Dawson-Hughes B, and Bischoff-Ferrari HA. (2004) Association between serum concentrations of 25-hydroxyvitamin D3 and periodontal disease in the US population. *Am. J. Clin. Nutr.*80:108-13.

Ding EL, Mehta S, Fawzi WW, and Giovannucci EL. (2007) Interaction of estrogen therapy with calcium and vitamin D supplementation on colorectal cancer risk: Reanalysis of Women's Health Initiative randomized trial. *Int. J. Cancer.* 122:1690-4.

Dobnig H, Pilz S, Scharnagl H, Renner W, Seelhorst U, Wellnitz B, et al. (2008) Independent Association of Low Serum 25-Hydroxyvitamin D and 1,25-Dihydroxyvitamin D Levels With All-Cause and Cardiovascular Mortality. *Arch. Intern .Med.*168:1340-9.

Dosil-Díaz O, Ruano-Ravina A, Gestal-Otero JJ, and Barros-Dios JM. (2007) Meat and fish consumption and risk of lung cancer: A case-control study in Galicia, Spain. *Cancer Lett.* 252:115-22.

Doughman SD, Krupanidhi S, and Sanjeevi CB. (2007) Omega-3 fatty acids for nutrition and medicine: considering microalgae oil as a vegetarian source of EPA and DHA. *Curr. Diabetes Rev.* 3:198-203.

East BR. (1939) Mean Annual Hours of Sunshine and the Incidence of Dental Caries. *Am. J. Public Health Nations Health.*29:777-80.

Eaton SB, Eaton SB 3rd, Sinclair AJ, Cordain L, and Mann NJ. (1998) Dietary intake of long-chain polyunsaturated fatty acids during the paleolithic. *World Rev. Nutr. Diet.*83:12-23.

Edwards MJ. (2007) Hyperthermia in utero due to maternal influenza is an environmental risk factor for schizophrenia. *Congenit Anom* (Kyoto).47:84-9.

Ehmann WD, Markesbery WR, Alauddin M, Hossain TI, and Brubaker EH. (1986) Brain trace elements in Alzheimer's disease. *Neurotoxicology.* 7:195-206.

Engeset D, Alsaker E, Lund E, Welch A, Khaw KT, Clavel-Chapelon F, et al. (2006) Fish consumption and breast cancer risk. The European Prospective Investigation into Cancer and Nutrition (EPIC). *Int. J. Cancer.* 119:175-82.

Farooqui AA, Horrocks LA, and Farooqui T. (2007) Modulation of inflammation in brain: a matter of fat. *J. Neurochem.* 101:577-99.

Farooqui AA, Ong WY, Horrocks LA, Chen P, and Farooqui T. (2007) Comparison of biochemical effects of statins and fish oil in brain: the battle of the titans. *Brain Res. Rev.* 56:443-71.

Fernandez E, Chatenoud L, La Vecchia C, Negri E, and Franceschi S. (1999) Fish consumption and cancer risk. *Am. J. Clin. Nutr.* 70:85-90.

Feskanich D, Singh V, Willett WC, and Colditz GA. (2002) Vitamin A intake and hip fractures among postmenopausal women. *JAMA.* 287:47-54.

Food and Agriculture Organization. (1996) *Food Balance Sheets*. Food and Agriculture Organization of the United Nations, Rome.

Foran JA, Good DH, Carpenter DO, Hamilton MC, Knuth BA, and Schwager SJ. (2005) Quantitative analysis of the benefits and risks of consuming farmed and wild salmon. *J. Nutr.* 135:2639-43.

Ford JS, and Myers RA. (2008) A global assessment of salmon aquaculture impacts on wild salmonids. *PLoS Biol.* 6:e33.

Forman JP, Giovannucci E, Holmes MD, Bischoff-Ferrari HA, Tworoger SS, Willett WC, et al. (2007) Plasma 25-hydroxyvitamin D levels and risk of incident hypertension. *Hypertension.*49:1063-9.

Fortes C, Mastroeni S, Melchi F, Pilla MA, Antonelli G, Camaioni D, et al. (2008) A protective effect of the Mediterranean diet for cutaneous melanoma. *Int. J. Epidemiol.* 37:1018-29.

Friesen EN, Ikonomou MG, Higgs DA, Ang KP, and Dubetz C. (2008) Use of terrestrial based lipids in aquaculture feeds and the effects on flesh organohalogen and fatty acid concentrations in farmed Atlantic salmon. *Environ. Sci .Technol.* 42:3519-23.

Fritschi L, Ambrosini GL, Kliewer EV, and Johnson KC; Canadian Cancer Registries (2004) Epidemiologic Research Group. Dietary fish intake and risk of leukaemia, multiple myeloma, and non-Hodgkin lymphoma. *Cancer Epidemiol. Biomarkers Prev.* 13:532-7.

Fukumoto K, Nagaoka I, Yamataka A, Kobayashi H, Yanai T, Kato Y, et al. (2005) Effect of antibacterial cathelicidin peptide CAP18/LL-37 on sepsis in neonatal rats. *Pediatr. Surg. Int.*21:20-4.

Gago-Dominguez M, Yuan JM, Sun CL, Lee HP, and Yu MC. (2003) Opposing effects of dietary n-3 and n-6 fatty acids on mammary carcinogenesis: The Singapore Chinese Health Study. *Br. J. Cancer.* 89:1686-92.

Garland CF, Garland FC, Gorham ED, Lipkin M, Newmark H, Mohr SB, et al. (2006) The role of vitamin D in cancer prevention. *Am. J. Public Health.*96:252-61.

Garland CF, and Garland FC. (1980) Do sunlight and vitamin D reduce the likelihood of colon cancer? *Int. J. Epidemiol.*9:227-31.

Garland CF, Gorham ED, Mohr SB, Grant WB, Giovannucci EL, Lipkin M, et al. (2007) Vitamin D and prevention of breast cancer: pooled analysis. *J. Steroid Biochem. Mol. Biol.* 103:708-11.

Geelen A, Schouten JM, Kamphuis C, Stam BE, Burema J, Renkema JM, et al. (2007) Fish consumption, n-3 fatty acids, and colorectal cancer: a meta-analysis of prospective cohort studies. *Am. J. Epidemiol.* 166:1116-25.

Ginde AA, Mansbach JM, and Camargo CA Jr. (2009) Vitamin D, respiratory infections, and asthma. Curr Allergy Asthma Rep. 9:81-7.

Giovannucci E, Liu Y, Rimm EB, Hollis BW, Fuchs CS. Stampfer MJ, et al. (2006) Prospective study of predictors of vitamin D status and cancer incidence and mortality in men. *JNCI.* 98:451-9.

Glatzel H. (1969) [Nutrition of the pregnant and nursing mother] *Z Allgemeinmed.*45:333-52. German.

Gochfeld M, and Burger J. (2005) Good fish/bad fish: a composite benefit-risk by dose curve. *Neurotoxicology*. 26:511-20.

Goldberg P, Fleming MC, and Picard EH. (1986) Multiple sclerosis: decreased relapse rate through dietary supplementation with calcium, magnesium and vitamin D. *Med. Hypotheses.*21:193-200.

Gorham ED, Garland CF, Garland FC, Grant WB, Mohr SB, Lipkin M, et al. (2007) Optimal vitamin D status for colorectal cancer prevention: a quantitative meta analysis. *Am. J. Prev. Med.* 32:210-6.

Grant WB. (1997) Dietary links to Alzheimer's disease. *Alz Dis Rev.*2:42-55. (http://www.sunarc.org/JAD97.pdf)

Grant WB. (1998) The APOE-epsilon4 allele and Alzheimer disease among African Americans, Hispanics, and whites. *JAMA.* 280:1662-3; author reply 1663.

Grant WB. (1999) Dietary links to Alzheimer's disease: 1999 update. *J. Alzheimers. Dis.* 1:197-201.

Grant WB. (2000) Fish consumption, cancer, and Alzheimer disease. *Am. J. Clin. Nutr.* 71:599-600.

Grant WB. (2002) An ecologic study of dietary and solar ultraviolet-B links to breast carcinoma mortality rates. *Cancer.* 94:272-81.

Grant WB, Campbell A, Itzhaki RF, and Savory J. (2002) The significance of environmental factors in the etiology of Alzheimer's disease. *J. Alzheimers Dis.* 4:179-89.

Grant WB, and Holick MF. (2005) Benefits and requirements of vitamin D for optimal health: a review. *Altern. Med. Rev.*10:94-111.

Grant WB, and Garland CF. (2006) The association of solar ultraviolet B (UVB) with reducing risk of cancer: multifactorial ecologic analysis of geographic variation in age-adjusted cancer mortality rates. *Anticancer Res.* 26:2687-99.

Grant WB. (2007) Vitamin D and Cancer Risk among American Indians. *Cancer Epidemiol. Biomarkers Prev.* 16:183.

Grant WB. (2009a) How strong is the evidence that solar ultraviolet B and vitamin D reduce the risk of cancer? An examination using Hill's criteria for causality. *Dermato-Endocrinology*. 1:17-24.

Grant WB. (2009b) Solar ultraviolet-B irradiance and vitamin D may reduce the risk of septicemia. *Dermato-Endocrinology*. 1:37-42.

Grant WB. (2009c) Does vitamin D reduce the risk of dementia? *J Alz Dis.* in press

Grant WB. (2009d) The health benefits of solar irradiance and vitamin D and the consequences of their deprivation. Ch. 39 in Holick MF (ed) Vitamin D - Physiology, Molecular Biology and Clinical Applications. Springer New Jersey, in press.

Grant WB, Cross HS, Garland CF, Gorham ED, Moan J, Peterlik M, et al. (2009) Estimated benefit of increased vitamin D status in reducing the economic burden of disease in Western Europe. *Prog Biophys Mol Biol.* In press.

Gupta R, Deedwania PC, Gupta A, Rastogi S, Panwar RB, and Kothari K. (2004) Prevalence of metabolic syndrome in an Indian urban population. *Int. J. Cardiol.* 97:257-61.

Gupta R, Gupta VP, Sarna M, Bhatnagar S, Thanvi J, Sharma V, et al. (2002) Prevalence of coronary heart disease and risk factors in an urban Indian population: Jaipur Heart Watch-2. Indian Heart J. 54:59-66.

Hall MN, Chavarro JE, Lee IM, Willett WC, and Ma J. (2008) A 22-year Prospective Study of Fish, n-3 Fatty Acid Intake, and Colorectal Cancer Risk in Men. *Cancer Epidemiol. Biomarkers Prev*. 17:1136-43.

Hamilton MC, Hites RA, Schwager SJ, Foran JA, Knuth BA, and Carpenter DO. (2005) Lipid composition and contaminants in farmed and wild salmon. *Environ. Sci. Technol.* 39:8622-9.

Hathcock JN, Shao A, Vieth R, and Heaney R. (2007) Risk assessment for vitamin D. *Am. J. Clin. Nutr.* 85:6-18.

Heaney RP, Davies KM, Chen TC, Holick MF, and Barger-Lux MJ. (2003) Human serum 25-hydroxycholecalciferol response to extended oral dosing with cholecalciferol. *Am. J. Clin. Nutr.* 77:204-10.

Hewison M, Burke F, Evans KN, Lammas DA, Sansom DM, Liu P, et al. (2007) Extra-renal 25-hydroxyvitamin D3-1alpha-hydroxylase in human health and disease. *J. Steroid Biochem. Mol. Biol.* 103:316-21.

Hewison M, Kantorovich V, Liker HR, Van Herle AJ, Cohan P, Zehnder D, et al. (2003) Vitamin D-mediated hypercalcemia in lymphoma: evidence for hormone production by tumor-adjacent macrophages. *J. Bone Miner. Res.* 18:579-82.

Hightower JM. (2009) *Diagnosis: Mercury; Money, Politics and Poison*. Island Press/Shearwater Books, Washington, DC, 277 pp.

Holick CN, Giovannucci EL, Stampfer MJ, and Michaud DS. (2006) A prospective study of fish, marine fatty acids, and bladder cancer risk among men and women (United States). Cancer Causes Control. 17:1163-73.

Holick MF. (2006) The role of vitamin D for bone health and fracture prevention. *Curr. Osteoporos. Rep.*4:96-102.

Holick MF. (2007) Vitamin D deficiency. N Engl J Med. 357:266-81.

Holick MF, Biancuzzo RM, Chen TC, Klein EK, Young A, Bibuld D, et al. (2008) Vitamin D2 is as effective as vitamin D3 in maintaining circulating concentrations of 25-hydroxyvitamin D. *J. Clin. Endocrinol. Metab.* 93:677-81.

Hollis BW. (2007) Vitamin D requirement during pregnancy and lactation. *J. Bone Miner. Res.* 22 Suppl 2:V39-44.

Hoogendijk WJ, Lips P, Dik MG, Deeg DJ, Beekman AT, and Penninx BW. (2008) Depression is associated with decreased 25-hydroxyvitamin D and increased parathyroid hormone levels in older adults. *Arch .Gen. Psychiatry*. 65:508-12.

Hope-Simpson RE. (1981) The role of season in the epidemiology of influenza. *J. Hyg.* (Lond).86:35-47.

Hosgood HD 3rd, Baris D, Zahm SH, Zheng T, and Cross AJ. (2007) Diet and risk of multiple myeloma in Connecticut women. *Cancer Causes Control.* 18:1065-76.

Houghton LA, and Vieth R. (2006) The case against ergocalciferol (vitamin D2) as a vitamin supplement. *Am. J. Clin. Nutr*. 84:694-7.

Huang TL, Zandi PP, Tucker KL, Fitzpatrick AL, Kuller LH, Fried LP, et al. (2005) Benefits of fatty fish on dementia risk are stronger for those without APOE epsilon4. *Neurology.* 65:1409-14.

Hylland K, Beyer J, Berntssen M, Klungsøyr J, Lang T, and Balk L. (2006) May organic pollutants affect fish populations in the North Sea? *J. Toxicol .Environ. Health* A. 69:125-38.

Ikonomou MG, Higgs DA, Gibbs M, Oakes J, Skura B, McKinley S, et al. (2007) Flesh quality of market-size farmed and wild British Columbia salmon. *Environ. Sci. Technol.* 41:437-43.

Ishida H, Seino Y, Matsukura S, Ikeda M, Yawata M, Yamashita G, et al. (1985) Diabetic osteopenia and circulating levels of vitamin D metabolites in type 2 (noninsulin-dependent) diabetes. *Metabolism.*34:797-801.

Jablonski NG, and Chaplin G. (2000) The evolution of human skin coloration. *J. Hum. Evol.* 39:57-106.

Jackson RD, LaCroix AZ, Gass M, Wallace RB, Robbins J, Lewis CE, et al. (2006) Calcium plus vitamin D supplementation and the risk of fractures. *N. Engl. J. Med.* 354:669-83.

Kaul N, Kreml R, Austria JA, Richard MN, Edel AL, Dibrov E, et al. (2008) A comparison of fish oil, flaxseed oil and hempseed oil supplementation on selected parameters of cardiovascular health in healthy volunteers. *J. Am. Coll Nutr.* 27:51-8.

Knekt P, Laaksonen M, Mattila C, Härkänen T, Marniemi J, Heliövaara M, et al. (2008) Serum Vitamin D and Subsequent Occurrence of Type 2 Diabetes. *Epidemiology.* 19:666-71.

Kornsteiner M, Singer I, and Elmadfa I. (2008) Very low n-3 long-chain polyunsaturated fatty acid status in Austrian vegetarians and vegans. *Ann. Nutr .Metab.*52:37-47.

Krajcovicova-Kudlackova M, Blazicek P, Mislanova C, Valachovicova M, Paukova V, and Spustova V. (2007) Nutritional determinants of plasma homocysteine. Bratisl Lek Listy.108:510-5.

Krkosek M, Ford JS, Morton A, Lele S, Myers RA, and Lewis MA. (2007) Declining wild salmon populations in relation to parasites from farm salmon. *Science.* 318:1772-5.

Kuhnlein HV, (2007) Receveur O. Local cultural animal food contributes high levels of nutrients for Arctic Canadian Indigenous adults and children. *J. Nutr.* 137:1110-4.

Kuhnlein HV, Receveur O, Soueida R, and Berti PR. (2008) Unique patterns of dietary adequacy in three cultures of Canadian Arctic indigenous peoples. *Public Health Nutr.* 11:349-60.

Lappe FM. (1971) *Diet for a Small Planet*. Ballentine Books, New York.

Lappe JM, Travers-Gustafson D, Davies KM, Recker RR, and Heaney RP. (2007) Vitamin D and calcium supplementation reduces cancer risk: results of a randomized trial. *Am. J. Clin. Nutr.* 85:1586-91.

Larsson SC, Håkanson N, Permert J, and Wolk A. (2006) Meat, fish, poultry and egg consumption in relation to risk of pancreatic cancer: a prospective study. *Int .J. Cancer.* 118:2866-70.

Lea E, and Worsley A. (2002) The cognitive contexts of beliefs about the healthiness of meat. *Public Health Nutr.* 5:37-45.

Lederman SA, Jones RL, Caldwell KL, Rauh V, Sheets SE, Tang D, et al. (2008) Relation between cord blood mercury levels and early child development in a World Trade Center cohort. *Environ Health Perspec*t. 116:1085-91.

Lee JH, O'Keefe JH, Lavie CJ, Marchioli R, and Harris WS. (2008) Omega-3 fatty acids for cardioprotection. *Mayo Clin. Proc.* 83:324-32.

Liu L, Howe P, Zhou YF, Xu ZQ, Hocart C, and Zhan R. (2000) Fatty acids and beta-carotene in australian purslane (Portulaca oleracea) varieties. *J. Chromatogr.* A. 893:207-13.

Liu PT, Stenger S, Tang DH, and Modlin RL. (2007) Cutting Edge: Vitamin D-mediated human antimicrobial activity against Mycobacterium tuberculosis is dependent on the induction of cathelicidin. *J. Immunol.* 179:2060-3.

Lu Z, Chen TC, Zhang A, Persons KS, Kohn N, Berkowitz R, et al. (2007) An evaluation of the vitamin D3 content in fish: Is the vitamin D content adequate to satisfy the dietary requirement for vitamin D? *J. Steroid .Biochem. Mol. Biol.* 103:642-4.

Luchsinger JA, Tang MX, Shea S, and Mayeux R. (2002) Caloric intake and the risk of Alzheimer disease. *Arch. Neurol.* 59:1258-63.

Lund B, Badskjaer J, Lund B, and Soerensen OH. (1978) Vitamin D and ischaemic heart disease. *Horm. Metab. Res.* 10:553-6.

Ma QL, Teter B, Ubeda OJ, Morihara T, Dhoot D, Nyby MD, et al. (2007) Omega-3 fatty acid docosahexaenoic acid increases SorLA/LR11, a sorting protein with reduced expression in sporadic Alzheimer's disease (AD): relevance to AD prevention. *J. Neurosci.* 27:14299-307.

MacLaughlin J, and Holick MF. (1985) Aging decreases the capacity of human skin to produce vitamin D3. *J. Clin. Invest.* 76:1536-8.

Mahaffey KR, Clickner RP, and Jeffries RA. (2009) Adult Women's Blood Mercury Concentrations Vary Regionally in the United States: Association with Patterns of Fish Consumption (NHANES 1999-2004). *Environ Health Perspect.* 117:47-53.

Markesbery WR. (1997) Oxidative stress hypothesis in Alzheimer's disease. *Free Radic. Biol. Med.*23(1):134-47.

Martins e Silva J. (2007) [Brief history of rickets and of the discovery of vitamin D] Acta Reumatol Port. 32:205-29. Portuguese.

McCullough ML, Robertson AS, Rodriguez C, Jacobs EJ, Chao A, Carolyn J, et al. (2003) Calcium, vitamin D, dairy products, and risk of colorectal cancer in the Cancer Prevention Study II Nutrition Cohort (United States). *Cancer Causes Control.* 14:1-12.

McGonigle RJ, Fowler MB, Timmis AB, Weston MJ, and Parsons V. (1984) Uremic cardiomyopathy: potential role of vitamin D and parathyroid hormone. *Nephron.*36:94-100.

McGrath J. (1999) Hypothesis: is low prenatal vitamin D a risk-modifying factor for schizophrenia? *Schizophr Res.*40:173-7.

Melamed ML, Michos ED, Post W, and Astor B. (2008) 25-hydroxyvitamin D levels and the risk of mortality in the general population. *Arch. Intern. Med.* 168:1629-37.

Melhus H, Michaëlsson K, Kindmark A, Bergström R, Holmberg L, Mallmin H, et al. (1998) Excessive dietary intake of vitamin A is associated with reduced bone mineral density and increased risk for hip fracture. *Ann. Intern. Med.* 129:770-8.

Miller GE, Grant PM, Kishore R, Steinkruger FJ, Rowland FS, and Guinn VP. (1972) Mercury concentrations in museum specimens of tuna and swordfish. *Science.* 175:1121-2.

Mookherjee N, Rehaume LM, and Hancock RE. (2007) Cathelicidins and functional analogues as antisepsis molecules. *Expert Opin. Ther. Targets.*11:993-1004.

Morris MC, Evans DA, Bienias JL, Tangney CC, Bennett DA, Wilson RS, et al. (2003) Consumption of fish and n-3 fatty acids and risk of incident Alzheimer disease. *Arch. Neurol.* 60:940-6.

Morris MC, Evans DA, Tangney CC, Bienias JL, and Wilson RS. (2005) Fish consumption and cognitive decline with age in a large community study. *Arch. Neurol.* 62:1849-53.

Mozaffarian D, and Rimm EB. (2006) Fish intake, contaminants, and human health: evaluating the risks and the benefits. *JAMA*. 296:1885-99.

Muhe L, Lulseged S, Mason KE, and Simoes EA. (1997) Case-control study of the role of nutritional rickets in the risk of developing pneumonia in Ethiopian children. *Lancet.*349:1801-4.

Munger KL, Levin LI, Hollis BW, Howard NS, and Ascherio A. (2006) Serum 25-hydroxyvitamin D levels and risk of multiple sclerosis. *JAMA.*296:2832-8.

Myers RA, Baum JK, Shepherd TD, Powers SP, and Peterson CH. (2007) Cascading effects of the loss of apex predatory sharks from a coastal ocean. *Science.* 315:1846-50.

Myers RA, and Worm B. (2003) Rapid worldwide depletion of predatory fish communities. *Nature.* 423:280-3.

Myers RA, and Worm B. (2005) Extinction, survival or recovery of large predatory fishes. *Philos. Trans R. Soc.Lond B Biol Sci.* 360:13-20.

Natri AM, Salo P, Vikstedt T, Palssa A, Huttunen M, Kärkkäinen MU, et al. (2006) Bread fortified with cholecalciferol increases the serum 25-hydroxyvitamin D concentration in women as effectively as a cholecalciferol supplement. *J. Nutr.* 136:123-7.

Norat T, Bingham S, Ferrari P, Slimani N, Jenab M, Mazuir M, et al. (2005) Meat, fish, and colorectal cancer risk: the European Prospective Investigation into cancer and nutrition. *J. Natl. Cancer Inst.* 97:906-16.

Novak EM, Dyer RA, and Innis SM. (2008) High dietary omega-6 fatty acids contribute to reduced docosahexaenoic acid in the developing brain and inhibit secondary neurite growth. *Brain Res.* 1237:136-45

Park KS, Nam JH, and Choi J. (2006) The short vitamin D receptor is associated with increased risk for generalized aggressive periodontitis. *J. Clin. Periodontol.*33:524-8.

Pham TM, Fujino Y, Kubo T, Ide R, Tokui N, Mizoue T, et al. (2008) Fish intake and the risk of fatal prostate cancer: findings from a cohort study in Japan. *Public Health Nutr.* (Epub ahead of print)

Philibert A, Vanier C, Abdelouahab N, Chan HM, and Mergler D. (2006) Fish intake and serum fatty acid profiles from freshwater fish. *Am. J. Clin. Nutr.* 84:1299-307.

Pilz S, Dobnig H, Fischer JE, Wellnitz B, Seelhorst U, Boehm BO, et al. (2008a) Low Vitamin D Levels Predict Stroke in Patients Referred to Coronary Angiography. *Stroke.* 39:2611-3.

Pilz S, März W, Wellnitz B, Seelhorst U, Fahrleitner-Pammer A, Dimai HP, et al. (2008b) Association of vitamin D deficiency with heart failure and sudden cardiac death in a large cross-sectional study of patients referred for coronary angiography. *J. Clin. Endocrinol. Metab.* 93:3927-35.

Pittas AG, Dawson-Hughes B, Li T, Van Dam RM, Willett WC, Manson JE, et al. (2006) Vitamin D and calcium intake in relation to type 2 diabetes in women. *Diabetes Care.*29:650-6.

Pittas AG, Lau J, Hu FB, and Dawson-Hughes B. (2007) The role of vitamin D and calcium in type 2 diabetes. A systematic review and meta-analysis. *J. Clin. Endocrinol. Metab.*92:2017-29.

Poon AH, Laprise C, Lemire M, Montpetit A, Sinnett D, Schurr E, et al. (2004) Association of vitamin D receptor genetic variants with susceptibility to asthma and atopy. *Am. J. Respir .Crit. Care Med.*170:967-73.

Porojnicu AC, Bruland OS, Aksnes L, Grant WB, and Moan J. (2008) Sun beds and cod liver oil as vitamin D sources. *J .Photochem. Photobiol. B.* 91:125-31.

Ramachandran A. (2005) Epidemiology of diabetes in India--three decades of research. *J. Assoc. Physicians India.* 53:34-8.

Rao S, Abdel-Reheem M, Bhella R, McCracken C, and Hildebrand D. (2008) Characteristics of High alpha-Linolenic Acid Accumulation in Seed Oils. *Lipids*. 43:749-55.

Rishton GM. (2008) Natural products as a robust source of new drugs and drug leads: past successes and present day issues. *Am. J. Cardiol.* 101:43D-49D.

Rollier A. (1927) *Heliotherapy, Second Edit.* Oxford Medical Publications. London.

Ros E, Núñez I, Pérez-Heras A, Serra M, Gilabert R, Casals E, et al. (2004) A walnut diet improves endothelial function in hypercholesterolemic subjects: a randomized crossover trial. *Circulation.* 109:1609-14.

Rostand SG. (1997) Ultraviolet light may contribute to geographic and racial blood pressure differences. *Hypertension*.30:150-6.

Roth DE, Jones AB, Prosser C, Robinson JL, and Vohra S. (2008) Vitamin D receptor polymorphisms and the risk of acute lower respiratory tract infection in early childhood. *J. Infect. Dis.*197:676-80.

Schaefer EJ, Bongard V, Beiser AS, Lamon-Fava S, Robins SJ, Au R, et al. (2006) Plasma phosphatidylcholine docosahexaenoic acid content and risk of dementia in Alzheimer disease. *Arch. Neurol.*63:1545–50.

Schellenberg D, Paré PD, Weir TD, Spinelli JJ, Walker BA, and Sandford AJ. (1998) Vitamin D binding protein variants and the risk of COPD. *Am. J. Respir. Crit. Care Med;*157:957-61.

Schwalfenberg, G, (2007) Not enough vitamin D: health consequences for Canadians. *Can. Fam. Physician;*53:841-54.

Schwalfenberg, G, (2008) Vitamin D and diabetes: improvement of glycemic control with vitamin D3 repletion. *Can. Fam. Physician*;54: 864-6.

Shaw SD, Brenner D, Berger ML, Carpenter DO, Hong CS, and Kannan K. (2006) PCBs, PCDD/Fs, and organochlorine pesticides in farmed Atlantic salmon from Maine, eastern Canada, and Norway, and wild salmon from Alaska. *Environ. Sci. Technol.* 40:5347-54.

Simopoulos AP, and Salem N Jr. (1986) Purslane: a terrestrial source of omega-3 fatty acids. *N. Engl. J. Med.* 315:833.

Simopoulos AP, Norman HA, Gillaspy JE, and Duke JA. (1992) Common purslane: a source of omega-3 fatty acids and antioxidants. *J. Am. Coll. Nutr.* 11:374-82.

Simopoulos AP. (1999a) Essential fatty acids in health and chronic disease. *Am. J. Clin. Nutr.* 70(3 Suppl):560S-569S.

Simopoulos AP. (1999b) Evolutionary aspects of omega-3 fatty acids in the food supply. Prostaglandins Leukot Essent Fatty Acids. 60:421-9.

Simopoulos AP. (2004a) Omega-3 fatty acids and antioxidants in edible wild plants. *Biol. Res.*37:263-77.

Simopoulos AP. (2004b) The traditional diet of Greece and cancer. *Eur. J. Cancer. Prev.* 13:219-30.

Simopoulos AP. (2008) The importance of the omega-6/omega-3 fatty acid ratio in cardiovascular disease and other chronic diseases. *Exp. Biol. Med. (Maywood)*. 233:674-88.

Singer P, Shapiro H, Theilla M, Anbar R, Singer J, and Cohen J. (2008) Anti-inflammatory properties of omega-3 fatty acids in critical illness: novel mechanisms and an integrative perspective. Intensive Care Med. 34:1580-92.

Smith MA, and Perry G. (1995) Free radical damage, iron, and Alzheimer's disease. *J. Neurol. Sci.* 134 Suppl:92-4.

Smith MA, Rottkamp CA, Nunomura A, Raina AK, and Perry G. (2000) Oxidative stress in Alzheimer's disease. *Biochim. Biophys. Acta.* 1502:139-44.

Sniadecki J, and Sniadecki J. (1840). On the cure for rickets Cited by Mozolowski W. Nature. (1909);143:121.

Stabler SP, and Allen RH. (2004) Vitamin B12 deficiency as a worldwide problem. *Annu. Rev. Nutr.*24:299-326.

Strumylaite L, Zickute J, Dudzevicius J, and Dregval L. (2006) Salt-preserved foods and risk of gastric cancer. *Medicina (*Kaunas).42:164-70.

Swann L. (2004) Fish nutrition and feeding. [Power Point slides) Retrieved from http://www.aquanic.org/courses/marine/slides/nutrition_2004.ppt#304,13,Conversion Efficiency

Takezaki T, Inoue M, Kataoka H, Ikeda S, Yoshida M, Ohashi Y, et al. (2003) Diet and lung cancer risk from a 14-year population-based prospective study in Japan: with special reference to fish consumption. *Nutr. Cancer.*45:160-7.

Terry P, Wolk A, Vainio H, and Weiderpass E. (2002) Fatty fish consumption lowers the risk of endometrial cancer: a nationwide case-control study in Sweden. *Cancer Epidemiol. Biomarkers Prev.* 11:143-5.

United States Department of Agriculture, Agricultural Research Service. (2007). USDA National Nutrient Database for Standard Reference, Release 20. Nutrient Data Laboratory Home Page, http://www.ars.usda.gov/ba/bhnrc/ndl

von Schacky C. (2007) Omega-3 fatty acids and cardiovascular disease. *Curr. Opin. Clin. Nutr. Metab. Care.* 10:129-35.

Wactawski-Wende J, Kotchen JM, Anderson GL, Assaf AR, Brunner RL, O'Sullivan MJ, et al. (2006) Calcium plus vitamin D supplementation and the risk of colorectal cancer. *N. Engl. J. Med*;354:684–96.

Wakai K, Tamakoshi K, Date C, Fukui M, Suzuki S, Lin Y, et al.; JACC Study Group. (2005) Dietary intakes of fat and fatty acids and risk of breast cancer: a prospective study in Japan. *Cancer Sci.* 96:590-9.

Wang TJ, Pencina MJ, Booth SL, Jacques PF, Ingelsson E, Lanier K, et al. (2008) Vitamin D deficiency and risk of cardiovascular disease. *Circulation*.117:503-11.

Webb AR. (2006) Who, what, where and when-influences on cutaneous vitamin D synthesis. *Prog. Biophys. Mol. Biol.* 92:17-25.

Webb AR, Kline L, and Holick MF. (1988) Influence of season and latitude on the cutaneous synthesis of vitamin D3: exposure to winter sunlight in Boston and Edmonton will not promote vitamin D3 synthesis in human skin. *J. Clin. Endocrinol .Metab.* 67:373-8.

Webb AR, and Engelsen O. (2006) Calculated ultraviolet exposure levels for a healthy vitamin D status. *Photochem. Photobiol.* 82:1697-703.

White L, Petrovitch H, Ross GW, Masaki KH, Abbott RD, Teng EL, et al. (1996) Prevalence of dementia in older Japanese-American men in Hawaii: The Honolulu-Asia Aging Study. *JAMA.* 276:955-60.

Wilkins CH, Sheline YI, Roe CM, Birge SJ, and Morris JC. (2006) Vitamin D deficiency is associated with low mood and worse cognitive performance in older adults. *Am. J. Geriatr. Psychiatry.* 14:1032-40.

Williams CM, and Burdge G. (2006) Long-chain n-3 PUFA: plant v. marine sources. *Proc. Nutr. Soc.* 65:42-50.

Wolk A, Larsson SC, Johansson JE, and Ekman P. (2006) Long-term fatty fish consumption and renal cell carcinoma incidence in women. *JAMA.* 296:1371-6.

World Health Organization. (2008) National health summaries. Retrieved October 18, 2008 from http://www.who.int/countries/en/.

Yang CX, Takezaki T, Hirose K, Inoue M, Huang XE, and Tajima K. (2003) Fish consumption and colorectal cancer: a case-reference study in Japan. *Eur. J .Cancer Prev.* 12:109-15.

Zatonski W, Campos H, and Willett W. (2008) Rapid declines in coronary heart disease mortality in Eastern Europe are associated with increased consumption of oils rich in alpha-linolenic acid. *Eur. J. Epidemiol.*23:3-10.

Zipitis CS, and Akobeng AK. (2008) Vitamin D supplementation in early childhood and risk of type 1 diabetes: a systematic review and meta-analysis. *Arch. Dis. Child.* 93:512-7.

Zittermann A, Schleithoff SS, Götting C, Dronow O, Fuchs U, Kuhn J, et al. (2008) Poor outcome in end-stage heart failure patients with low circulating calcitriol levels. *Eur. J. Heart Fail.*10:321-7.

In: Fish Consumption and Health
Editors: George P. Gagne and Richard H. Medrano
ISBN 978-1-60741-151-2

Chapter III

Differences in Understanding of the Risks and Benefits of Fish Consumption in Three Coastal Estuaries in New York and New Jersey

Joanna Burger
Division of Life Sciences, Rutgers University 604 Allison Road, Piscataway, New Jersey, USA

Abstract

To understand consumption patterns for self-caught fish it is essential to examine the knowledge base used by people in making decisions, including both the risks and benefits. Public perceptions of warnings about contaminants in fish have been frequently examined from a number of different locations, but seldom have both the risks and benefits been examined together, especially from different nearby locations. In this chapter I examine perceptions of the risks, warnings about, and benefits of fish consumption in three coastal estuaries. Fishermen and other recreationists (N = 329) were interviewed in Barnegat Bay New Jersey, the New York/New Jersey Harbor, and along the northern shore of Long Island in 2007 and 2008. Overall, a higher percentage of people from all three regions had heard about the benefits of fish consumption compared to the risks (or warnings about fish consumption). While there were no locational differences in having heard about the risks from fish consumption or in belief in any warnings, there were significant differences in knowledge about warnings, differences in contaminant loads among fish, and health benefits from eating fish. People from Long Island were less aware of warnings about fish consumption, and fewer knew that contaminants loads differed among fish, although the fewest people from the harbor were aware of the benefits of fish consumption. There were several other major differences: 1) awareness of mercury was higher in people interviewed from Barnegat Bay compared to elsewhere, 2) Word of mouth was they method over 55 % of the people from Barnegat Bay heard about warnings compared to 13 % from the harbor (and none on Long Island), 3) More people from Long Island had heard about warnings for bluefish and striped bass compared to those from elsewhere, 4) A higher percentage of people from the harbor knew about contaminants in bottom-dwelling fish compared to elsewhere, and 5) Fewer people from the harbor knew that fish were generally healthy and had high Omega-3 oils. These data illustrate that even people from the same region do not have the same knowledge base and that benefits of fish consumption are more widely known generally than the risks, although specific information about benefits is generally low. Public policy makers, risk managers, and health professionals need to be aware of the knowledge base of local residents, and a targeted information campaign is necessary to more broadly inform the public about the specific warnings (relative to both contaminants and fish species) and health benefits of fish consumption so that they can make informed decisions about consumption.

Introduction

Health professionals, risk assessors, and environmental managers are interested in reducing the risk to humans and other receptors from contaminants in the environment For environmental managers and regulators, this takes the form of reducing the levels of heavy metals and other contaminants in the environment, either by remediation or reduction at the source. One of the pressing public health issues is the possibility of disease or developmental abnormalities from the consumption of food, especially fish with high levels of contaminants, such as mercury and PCBs..

Fish consumption is a classic case of consumer choice. In their everyday lives, people must choose what type of foods to eat. Most protein sources have benefits and risks, and these must be weighed against those of other types of proteins. The choices people face include what type of protein to eat (fish, poultry, red meat, plant protein), how much of each to eat, how often to eat it and, in the case of fish, what kinds and sizes to eat. For many fish consumers, the choices include what kinds of fishing to do (where and how), what fish to target, what kind and size of fish to keep for consumption, what part of the fish to prepare for cooking (whole, fillet), and how to cook the fish (fry, broil, bake). All these choices influence both the risks and benefits from fish consumption. Consuming fish and shellfish thus involves risk balancing (comparing alternative risks and benefits; see papers in Am J. Prev. Med., Vol. 29, 2005). Yet risk managers, public policy makers, and environmental managers cannot design informational material without understanding what the public knows and understands about the risks and benefits of fish consumption.

Fish are a good source of low-fat protein that provides many health benefits, such as omega-3 (n-3) fatty acids that reduce cholesterol levels and the incidence of stroke and heart disease in adults, reduce pre-term deliveries, and enhance cognitive development in fetuses and young children (Bouzan et al, 2005; Cohen et al, 2005a,b; Daviglus et al, 2002; Konig et al, 2005; McMichael and Butler2005; Patterson 2002; Willett 2005). However, contaminants, such as polychlorinated biphenyls (PCBs) and mercury, are sufficiently high in some fish and seafood to pose a potential health risk to consumers (Gochfeld 2003; Hightower and Moore 2003; IOM 2006; NRC 2000). Methylmercury (MeHg) counteracts the cardioprotective effects of omega-3 fatty acids (Guallar et al, 2002). Fish consumption is the most significant source of methylmercury exposure for the public (Rice et al, 2000), for example, and poses a risk to susceptible adults, fetuses, neonates, and developing infants and young children

States respond to the potential risk from contaminants in fish by issuing state advisories for self-caught fish (usually freshwater), and rarely do they address consumption of commercial fish. The U.S. Food and Drug Administration (FDA) has issued advisories for four species of commercial fish (Burger and Gochfeld 2004; FDA 2001, 2003, 2005, 2006), but little attention has been devoted to self-caught saltwater fish, or to the knowledge base of people fishing in coastal waters, despite the importance of recreational fishing to the economies of coastal states.

Moreover, fishing is a popular pastime, and for many it provides a way to relax with family and friends, and commune with nature (Burger 2004). It also provides protein not otherwise available (Burger 2002; Burger and Gochfeld 2006; Knuth et al, 2003; Toth and Brown 1997). Fishing is important in rural regions, for Native Americans, for subsistence people (whether Native Americans or not), and for recreationists (Burger 2004; Burger et al, 1999a,b, 2001, 2007; Dellinger 2007; Hamrick and Smith 2003; Harper and Harris, in press; Harris and Harper 1997, 2000; Ramos and Crain 2001; Silver et al, 2007; Stumpff 2006). The key factor for risk assessors, risk managers, and public health professionals is how much fish people eat, and thus who is at risk, as well as the knowledge base they use to make informed decisions about fish consumption. Thus, it is not only subsistence people who are at risk, but any person who consumes large quantities of fish with high levels of mercury, PCBs or other contaminants.

In this chapter, the knowledge and perceptions of people fishing and recreating in three coastal regions of the New York Bight (Barnegat Bay in New Jersey, the New York/New Jersey harbor, and Long Island) are examined regarding their general and specific knowledge of the risks and benefits of fish consumption. Knowledge of both risks and benefits is essential for individuals to make informed decisions about fish consumption, and to select particular fish. While many studies have examined perceptions about warnings, they have not examined either specific warnings or specific benefits, nor have they examined local differences that might affect decisions. The distinction was made between warnings and knowledge about the risks from fish consumption because it was essential to determine whether people translated warnings into an understanding that these posed a health risk.

This paper particularly addresses geographical differences in the knowledge base (see NYSDOH 2005, Burger and Gochfeld 2008, for a more general treatment of the findings for the region). Understanding locational differences is particularly important for the New York Bight because the region is heavily populated, some coastal areas are important tourist and recreational areas (Barnegat Bay), the region receives immigrant waves that make frequent information campaigns necessary (especially for the NY/NJ harbor), and targeted informational campaigns may be necessary for large coastal regions. The data presented in this chapter show that even within a limited regional area, there are knowledge-based differences that affect individual choices, and that therefore require targeted informational campaigns by health professionals, risk assessors, and environmental managers. The findings have implications for other large coastal bays and estuaries, and suggest that site-specific information is required to adequately design a program of risk management. A key element of risk management is supplying the public with appropriate, site-specific information to make informed decisions. Yet, the efficacy of such communication programs is seldom examined.

Methods

The overall protocol was to interview people fishing or otherwise engaged in recreation in three coastal areas: Barnegat Bay, New Jersey, the New York/New Jersey harbor, and the north shore of Long Island (Figure 1). All interviewers were trained, had practice sessions, and had interviewed people using the same questionnaire in previous studies (this study included a Spanish-speaking interviewer). Parks, fishing piers, beaches, and other open-access places along the shore were visited regularly, on weekdays and on weekends, at all times of the day (from 6 AM until 9 PM). The interviewers identified themselves from Rutgers University and asked then permission to interview them. Respondents were assured that their identity was not being recorded, and that no individual identifiers were obtained. Only 8 % of those approached refused to be interviewed, largely because they had small children, were leaving shortly, were talking to friends, or the fishing was so good they could not take time away to be interviewed.

The questions discussed in this chapter were embedded within a larger questionnaire that included questions about fish consumption, fishing behavior, fish populations, and natural resource damage assessment. Open-ended questions relevant to the present study included: Do you think there are risks to eating fish (and if so, what)? Have you heard any warnings about eating fish (if so, what)? Do you believe the warnings? Do you think there are benefits to eating fish (and if so, what); and Do you think the fish are contaminated (and if so, what species, with what contaminant)? Demographic information (age, gender, educational level, household income, and self-identified ethnicity) was asked at the end of the questionnaire, and people were told that they did not have to answer. We recorded all their answers; respondents could give more than one response to any question. The survey took from 20 min to 40 min, depending upon how many questions people asked. This study was approved as exempt by the Rutgers University Institutional Review Board because it was non-invasive and subjects were not identified in the data.

Non-parametric one-way analysis of variance was performed using the Statistical Analysis System (PROC NPAR1WAY with the Wilcoxon option, SAS 1995), which yields a chi square statistic. A probability level of 0.05 was considered significant.

Results

Population Characteristics

Of the 329 people interviewed, 56 % were college graduates, 84 % were white, 82 % were male, the average age was 46 years (SE = 0.9), and the average household income was \$80,900 (SE = \$3,500). Although there were no locational differences in ethnicity and mean age, there were significant differences in education ($\chi^2 = 18.3$, $P < 0.001$) and income ($X^2 =$

17.4, $P < 0.0002$; Table 1). Interviewees living in the NY/NJ harbor included fewer college graduates and lower mean incomes than those from elsewhere. There was an interaction between education and income, but not age and income (Table 2); people with more education had higher incomes.

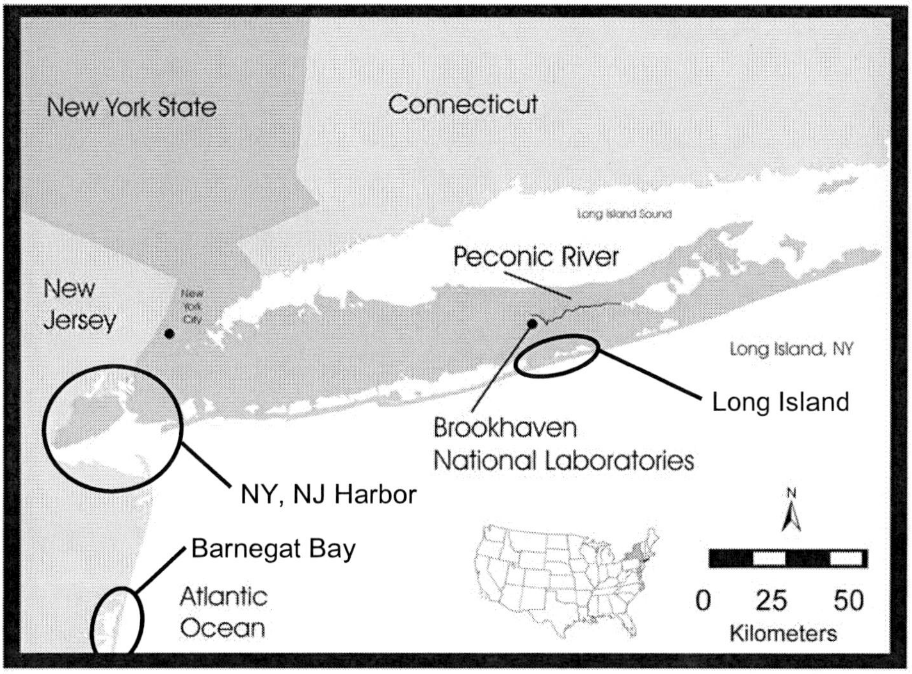

Figure 1.

Table 1. Demographics of study population in New Jersey and New York (2007). Given are means ± standard error for age and income

	Overall	Barnegat Bay	NY, NJ Harbor	Long Island	X^2
	N=329	N=123	N=103	N=103	
Education					18.3 (0.001)
Less than high school	20%	26%	24%	8%	
High school graduate	24%	16%	25%	33%	
College graduate	56%	58%	51%	59%	
Ethnicity					10.5 (NS)
White	84%	88%	77%	84%	
Black	5%	4%	8%	2%	
Hispanic	4%	2%	8%	4%	
Other/Unidentified	7%	6%	7%	10%	
Age	46.0 ± 0.9	45.0 ± 1.5	45.8 ± 1.7	47.1 ± 1.6	1.3 (NS)

Table 1. (Continued)

	Overall	Barnegat Bay	NY, NJ Harbor	Long Island	X^2
	N=329	N=123	N=103	N=103	
Income (thousands of dollars)	80.9 ± 3.5	86.5 ± 6.2	65.7 ± 5.9	88.9 ± 5.4	17.4 (0.0002)
Gender (% female)	18%	16%	14%	25%	

Table 2. Demographics of study population in New Jersey and New York (2007). Given are means ± standard error and range for age and income (N=329)

	Percent	Income (thousands of dollars)	Age	Gender (%Women)	Ethnicity (% White)
Overall		80.9 ± 3.5, 0-500	46.0 ± 0.9, 13-90	18%	84%
Education					
Less than high school	20%	61.6[a] ± 8.7, 25-100	48.9 ± 2.3, 13-87	12%	88%
High school graduate	24%	61.2 ± 4.3, 0-150	46.8 ± 2.0, 15-90	16%	73%
Some college	17%	77.5 ± 5.2, 30-150	47.2 ± 2.2, 20-74	33%	87%
College graduate	30%	87.8 ± 4.6, 20-250	43.5 ± 1.6, 19-75	14%	88%
Graduate level education	9%	109 ± 17.0, 0-500	43.5 ± 2.0, 21-60	19%	84%
X^2		19.4 (0.0007)	5.5 (NS)	8.6 (NS)	1.7 (NS)

a. only 10 of 65 responded.

Consumption Warnings and Benefits of Fish Consumption

The general knowledge base differed significantly among the three study sites (Tables 3 and 4). Overall, there were significant differences in awareness of warnings, the exact nature of the warnings, awareness of differences as a function of species of fish, the contaminants that pose a risk, the benefits of fish consumption, and whether they would like more information about the safety of fish consumption (Figure 2).

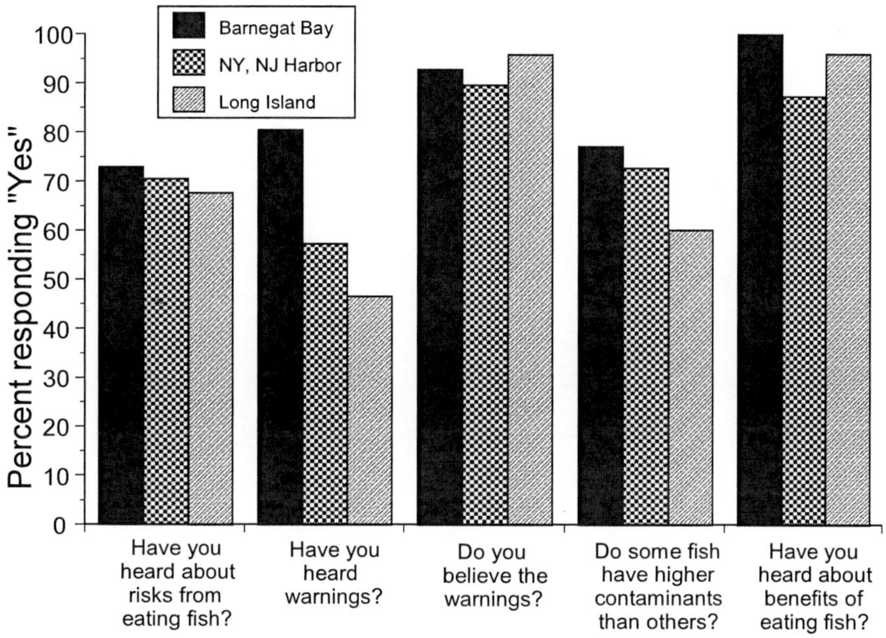

Figure 2.

In general, more people from Barnegat Bay knew about all aspects of the questions than did those fro elsewhere, and people from the NY-NJ harbor were less aware of warnings (event though there were more from that region) and were less aware that there were differences among fish with respect to risk.

The specific knowledge base also differed significantly among the locals with respect to contaminants of concern, sources of information, species of special concern, specific health benefits, and specific risks. Nearly twice as many people from Barnegat Bay knew about warnings with respect to mercury, Overall, three times as many people knew about the risk of mercury compared to the risks from PCBs, and less than 5 % could name any other contaminant (or bioagent) of concern (Tables 3 and 4). Despite the relatively high awareness of consumption advisories, few people in each area could attribute the risks to particular contaminants (Table 4).

The source of information about fish consumption warnings differed among regions, and these differences are useful for local risk managers and health professionals (Table 3). People from Barnegat Bay were aware of obtaining information on warnings from many different sources, while far fewer from Long Island obtained information from a number of sources. Interestingly, word of mouth was far more important for people interviewed in Barnegat Bay than elsewhere. Interestingly, for the people interviewed, the internet played almost no role in education with respect to fish consumption.

Although a relatively high percentage of people from all three of the study sites recognized that there were differences in contaminants in different species of fish, when asked directly which fish have relevant warnings, less than 60 % could list the fish species of concern. Few people mentioned that bigger fish have higher levels of contaminants, even though this is a well-accepted concept.

Table 3. Knowledge of fish warnings by location. Percent of people responding to open ended questions

	Barnegat Bay	NY, NJ Harbor	Long Island	X^2
	N=123	N=103	N=103	
Heard any warnings for fish? (Percent answering "Yes")	80.5%	57.3%	46.6%	28.9 (<0.0001)
What warning? (Answers for those responding yes)				45.9 (0.0003)
Mercury	69.1%	33.0%	37.9%	
PCB's	20.3%	17.5%	14.6%	
Don't Know	5.7%	10.7%	1.9%	
Pollution	0.0%	4.9%	1.0%	
Bacteria	0.0%	0.0%	5.8%	
Dioxins	0.8%	1.9%	0.0%	
Lead	0.0%	1.9%	1.0%	
Selenium	1.6%	0.0%	0.0%	
Other	8.1%	5.8%	2.9%	
Source of Warnings?				97.4 (<0.0001)
Newspaper	38.2%	28.2%	20.4%	
Word of Mouth	56.1%	12.6%	0.0%	
Magazine	37.4%	12.6%	19.4%	
DEP	22.8%	7.8%	12.6%	
TV	12.2%	23.3%	5.8%	
Sign	15.5%	16.5%	5.8%	
Radio	9.8%	2.9%	2.9%	
Internet	0.0%	3.9%	1.0%	
Do you Believe the Warnings? (Percent answering "Yes")	92.9%	89.7%	95.9%	5.1 (NS)
Are some fish worse in contaminants? (Percent answering "Yes")	77.2%	72.8%	60.2%	21.8 (0.0002)
What fish? (Answers for those responding yes)				119.5 (<0.0001)
Bluefish	43.2%	28.0%	56.5%	
Striped Bass	23.2%	17.3%	27.4%	
Bottom Fish	3.2%	22.7%	8.1%	

	Barnegat Bay	NY, NJ Harbor	Long Island	X^2
	N=123	N=103	N=103	
Shark	2.1%	9.3%	17.7%	
Tuna	1.0%	9.3%	16.1%	
Don't Know	3.2%	10.7%	3.2%	
Bigger fish	4.2%	4.0%	4.8%	
Depends on location	6.3%	4.0%	0.0%	
All Fish are bad	1.0%	1.3%	0.0%	
Other	8.4%	5.3%	1.6%	
Heard any benefits for fish? (Percent answering "Yes")	100.0%	86.4%	96.1%	16.8 (0.0002)
What benefits? (Answers for those responding yes)				61.6 (<0.0001)
Generally healthy	45.9%	18.5%	23.3%	
High in Omega 3	23.8%	20.4%	44.7%	
Fish oils are healthy	17.2%	12.6%	18.5%	
Lowers cholesterol	5.0%	9.7%	1.0%	
Brain food	0.8%	9.7%	3.9%	
High in protein	2.5%	5.8%	1.0%	
Good for your heart	0.0%	1.9%	1.9%	
Better than red meat	0.8%	1.9%	0.0%	
Don't know	3.3%	4.9%	0.0%	
Other	0.8%	1.0%	1.9%	

Surprisingly, although over 85 % of people interviewed knew there were health benefits from eating fish, people were mainly aware of that fish had healthy fish oils. Less than 10 % said that fish contribute to lowering cholesterol, and less than 6 % said that fish were high in protein (Table 3).

Even with the lack of information about specific details of chemicals of concern, chemicals that pose a risk, and fish species with contaminants of concern, most people said they did not need more information. Only 7 to 14 % of the people that were interviewed said they wanted to know which fish are safe to eat, and 3 to 23 % (depending upon locale), wanted to know how much is safe to eat. Very few people (0, 0 and 9.6 %) wanted to know if the data were current, a very sophisticated question that is very important (given that many metals are decreasing in the environment).

Overall, there were fewer geographical differences in the information base about risks, than about the warnings (or advisories); a higher percentage of people from Barnegat Bay were aware of the warnings than elsewhere (Figure 2). While most people believe the warnings, and recognize that some fish have higher levels than others, fewer interviewees knew which fish were the most contaminated. Finally, over 85 % of people interviewed in all three study areas (Figure 2). This indicates that the media has done a better job of making people aware of the benefits of consuming fish.

Discussion

Risk communication is clearly an important factor in reducing the risk from fish consumption, particularly for mercury, and the message must clearly give information on species and sizes of fish with the highest contaminant levels (Lange et al, 1994, Anderson et al, 2004, Burger 2005). The information, however, is more complex than initially believed, and people must be supplied with information on a number of different risk balancing issues (risks and benefits of fish consumption, and of alternative protein sources, see below). Further, information on the efficacy of communication with respect to fishing, fish consumption and risk should be formalized by the use of recognized indicators (Burger 2008). That is, as risk assessors, perceptions scientists, and health professionals, we need to ask the same questions so that we can begin to form a synthesis across geographical, ethnic, cultural, and economic regions. For too long different states and regions have conducted studies with such different objectives that the results cannot be compared. More serious is the fact that consumption patterns cannot be generalized across regions, ethnic groups, or recreational groups because the same questions are not asked uniformly.

Hazards, Risks and Benefits of Fish Consumption

In this chapter a distinction is made between understanding of warnings, and understanding of risks and benefits of fish consumption. Warnings are issued by states, and in some cases, by the FDA (2001, 2003, 2005, 2006). These warnings are issued based on contaminant levels in fish and fish consumption patterns; thus they are loosely based on potential risk, often to the most sensitive target audience (although this may not reflect the high-end, subsistence, or Native American consumers). Presumably people who follow the consumption warnings (or advisories) issued by states will not be unduly exposed to chemicals.

In contrast, people need to balance the benefits of fish consumption with the risks of eating fish, of not eating fish, and of eating some other form of protein (or no or little protein). That is, people prefer to eat a certain amount of protein (and need a certain amount), and must decide what foods to eat that will provide that protein. Health conscious people often choose fish because it is perceived as a healthier form of protein than red or fatty meat. Several other factors enter their decision, such as taste and preference, cost, availability, and health objectives. For many people, eating fish is part of the pleasure of fishing and outdoor recreation (Toth and Brown 1997).

Another distinction needs to be made: knowing about warnings relates directly to awareness of specific warnings issued for local areas, while knowing about potential risks from fish consumption is a more general knowledge base, which could be gained from health professionals, family and friends, and the national media, among others. Warnings about fish consumption usually are posted in local waters, are described in literature that accompanies fishing licenses, and are reported in newspapers and other media.

The results from this study indicate that a higher percentage of people knew about the risks of fish consumption than knew about warnings or advisories, but the highest percentage of people knew about the benefits of fish consumption, a subject that has increased in prominence in the media and from health professionals. Most people who knew about warnings believed them, but still only about 60 % of people knew about the warnings. Further, less than half the people interviewed actually knew what species of fish were covered in the warnings, and less than 5 % knew about the warnings issued by the FDA (for tuna and shark, for example). That is, a general awareness of warnings without knowing the specifics will not allow people to make informed decisions. If people do not know what kinds of fish are under advisories or warnings, they cannot reasonably avoid eating them (or eating too many meals of that fish). Thus, merely knowing there are warnings does not help them make a health-related decision.

An important point that emerges from the data is the need to educate the public about the importance of understanding: 1) local, regional and national warnings about contaminants in fish that relate to the fish the eat, 2) specific information about the warnings, 3) risks from consuming fish (amount of fish, species of fish, and sizes of fish), 4) benefits of fish consumption, and 5) how to begin balancing the risks and benefits of fish consumption, with the risks and benefits of not consuming fish (and consuming other sources of protein). This is a complicated task, and one that needs to be addressed holistically, rather than the usual method of different agencies dealing with parts of the problem. State agencies, such as environmental protection, tourism and recreation, and public health, should meet regularly to develop a joint strategy for the provision of information to the public, environmental managers, and health professionals.

Locational Differences in Information Base

The main objective of this chapter was to determine if there are differences in the information base of people fishing and consuming fish in three coastal estuaries.in the New York Bight. This question is important for the region because it relates directly to both risk management and the development of educational and outreach materials for the local agencies, but more generally because it illustrates both the kinds of information needed by health professionals, risk assessors, and risk managers, and the potential for significant differences between nearby local areas. Such differences have important implications for designing not only warnings about fish consumption (produced by state agencies), but for health professionals who must provide information to fish consumers, particularly pregnant women (women of child-bearing age), infants and children.

Despite the geographical closeness of the study areas, and the fact that all are in the same media range (TV, radio and newspapers such as the *NY Times*), there were significant differences in awareness of warnings, the exact nature of the warnings, awareness of differences as a function of species of fish (and the fish involved), the types of risks from fish consumption (but not that there were risks), the benefits of fish consumption, and whether they would like more information about the safety of fish consumption. This was surprising, given the fact that the whole region is served by the same media (TV, radio, some newspapers), and that there are consumption advisories (or warnings) for all three areas for mercury and PCBs). For all the major questions, a higher percentage of the people in Barnegat Bay had greater knowledge than those from elsewhere (refer to Figure 2), and the differences were significant for awareness of warnings, knowledge that some fish have higher levels of contaminants, and the benefits of fish consumption. Surprisingly, a significantly lower percentage of people from Long Island, interviewed near the Brookhaven National Laboratory where there are state-issued mercury advisories, had knowledge about warnings, and about differences in contaminants among fish species, that would allow them to make informed decisions about what species of fish to eat.

Although there was little disparity between the percentage of people who had heard warnings and knew there were risks from fish consumption (Figure 2), there was disconnect between knowledge of warnings, and specific knowledge of risks. For example, between 33 and 69 % of people were aware there were warnings about fish consumption with respect to mercury, but only 10 to 30 % of people thought there were risks from mercury in fish. This is remarkable because it suggests that people can be aware of the warnings, but not understand that eating fish with high mercury means there are health risks. A similar disparity occurred for PCBs. Although this difference may simply reflect that people do not understand the meaning of risk, it may run deeper and reflect individual bias that they personally are not at risk (though others in the same situation may be).

The differences were not consistent across the study sites. For example, the people interviewed in Barnegat Bay exhibited the highest percentage of people who knew about warnings with respect to mercury ((69 %), but the lowest percentage (about 10 %) said mercury posed a risk. In contrast, only 33 % of people interviewed in the NY.NJ harbor knew about mercury warnings, and 34 % knew that there were risks from mercury when eating fish. Further, the percent of people who mentioned that pregnant women and children are at risk was relatively small, but differed among regions. Over 30 % of people from Long Island were aware of these risks, while it was only 1 % of those interviewed from the NY-NJ harbor. This is remarkable, since it is often pregnant women that are the target of media messages, and presumably, for health professionals.

There are two major aspects of the results that require discussion: 1) there are major differences in the awareness of advisories and knowledge of the risks and benefits of fish consumption, and, 2) there are major differences in the disparity between specific knowledge about the contaminants leading to the advisories, what contaminants pose a risk, who is at risk, and what the specific health benefits are. Both of these aspects need to be addressed by local agencies, as well as by regional (states and EPA), and national agencies (e.g. FDA). The existence of significant differences in awareness and knowledge about advisories, risks and benefits of fish consumption in nearby estuaries suggests that site-specific information on awareness and knowledge is required to truly address the needs of local residents, particularly in places where the major or favorite fish species are under advisories (e.g. bluefish and striped bass for the New York Bight).

Conclusions

Health professionals, risk assessors, and environmental managers are interested in reducing the risk to humans and other receptors from contaminants in the environment. One of the pressing public health issues is the possibility of disease or developmental abnormalities from the consumption of food, especially fish with high levels of contaminants. Public health officials, health professionals, and environmental managers can inform the public of the possible risks by developing fish consumption advisories that describe the fish that should be limited or avoided, and health professionals can develop brochures and other informational materials that can be used to inform the public. Further, health professionals can discuss the risks and benefits of consumption with their patients directly, allowing the opportunity for questions. This latter mechanisms requires further education of a range of health professionals.

Such communication mechanisms, however, are only useful if people read and digest the information, and make informed decisions about their choices (whether to eat fish, what species to eat, how often to eat fish). The data in this paper suggests that there are differences in the knowledge base about both warnings and risk from fish consumption in bays that are within the same media region. Specific differences were that: 1) awareness of mercury was higher in people interviewed from Barnegat Bay compared to elsewhere, 2) Word of mouth was they method over 55 % of the people from Barnegat Bay heard about warnings compared to 13 % from the harbor (and none on Long Island), 3) More people from Long Island had heard about warnings for bluefish and striped bass compared to those from elsewhere, 4) A higher percentage of people from the harbor knew about contaminants in bottom-dwelling fish compared to elsewhere, and 5) Fewer people from the harbor knew that fish were generally healthy and had high Omega-3 oils. This study clearly shows that even people from the same region do not have the same knowledge base, and that benefits of fish consumption are more widely known generally than the risks, although specific information about benefits is generally low. A detailed plan is need for regions, such as the New York- New Jersey harbor.

More generally, however, the results suggest that as a Nation, we need to devote more time to formalizing both our method of collecting information on fish consumption, knowledge of the risks and benefits of fish consumption, and adherence to such information, and to our methods of informing the public about such risks and benefits. Standardization of information acquisition will help to develop guidance for dealing with the problem, and can lead to a National program to understand how people make decisions about fish consumption (and other food choices), Such information will become even more critical as we move toward the more global production and transport of food.

Acknowledgments

I thank the many people who have contributed to various aspects of my research and thinking about fish consumption and ecosystem health, including M. Gochfeld, B.D. Goldstein, C.W. Powers, C. Chess, and K. Cooper, and others have contributed logistical support, including C. Jeitner, S. Shukla, and M. Donio. This research was conducted under approved Rutgers University protocols, and was partly funded by Division of Science, Research and Technology, New Jersey Department of Environmental Protection, the Jersey Coast Angler's Association, a National Institute of Environmental Health Sciences Center (P30ES005022), a Hatch grant, and the Consortium for Risk Evaluation with Stakeholder Participation (DE-FC01-06EW07053). The results and ideas expressed in this chapter are the author's, and do not necessarily reflect those of any of the funding agencies.

References

Anderson, H.A., Hanrahan, L.P., Smith, A., Draheim, L., Kanarek, M, and Olsen,. J. 2004. The role of sport-fish consumption advisories in mercury risk communication: a 1998-1999 12-state survey of women age 18-45. *Environ. Res. 95,* 315-324.

Bouzan, C., Cohen, J.T., Connor, W.E., Kris-Etherton, P.M., Gray, G.M., Konis, A., Lawrence, R.S., Savitz, D.A, and Teutsch, S.M. 2005. A quantitative analysis of fish consumption and stroke risk. *Am. J. Prev. Med. 29*, 347-352.

Burger, J. 2002. Consumption patterns and why people fish. Environ. Res. 90, 125-135.

Burger, J. 2004. Fish consumption advisories: knowledge, compliance and why people fish in an urban estuary. J. Risk Res. 7, 463-479.

Burger, J. 2005. Fishing, fish consumption, and knowledge about advisories in college students and others in central New Jersey. Environ. Res. 98, 268-275.

Burger, J. 2008. Perceptions as indicators of potential risk from fish consumption and health of fish populations. *Environ. Indicators* (in press)

Burger, J.and Gochfeld, M. 2004. Mercury in canned tuna: white versus light and temporal variation. *Environ. Res. 94*, 239-49.

Burger, J. and Gochfeld, M. 2006. A framework and information needs for the management of the risks from consumption of self-caught fish. *Environ. Res. 101*, 275-285.

Burger, J. and Gochfeld, M. 2008. Mercury in fish caught by New Jersey coastal anglers. *JCAA Newletter*, April 2008. 3 pages. In press

Burger, J., Pflugh, K.K., Lurig, L., vonHagen, L.A. and vonHagen, S.A.. 1999a. Fishing in urban New Jersey: ethnicity affects information sources, perceptions, and compliance. *Risk. Anal. 19*, 217-229.

Burger, J., Stephens, C.S., Boring, S.C., Kuklinski, M., Gibbons, J.W., and Gochfeld, M. 1999b. Factors in exposure assessment: ethnic and socioeconomic differences in fishing and consumption of fish caught along the Savannah River. *Risk Anal. 19*, 217-229.

Burger, J., Gaines, K.F., and Gochfeld, M. 2001. Ethnic differences in risk from mercury among Savannah River fishermen. *Risk Anal. 21,* 533-544.

Burger, J., Gochfeld, M., Jeitner, C., Burke, S., Stamm, T., Snigaroff, R., Snigaroff, D., Patrick, R., and Weston, J. 2007. Mercury levels and potential risk from subsistence foods from the Aleutians. Sci. Total Environ. 384, 93-105.

Cohen, J.T., Bellinger, D.C. and Shaywitz, B.A.. 2005a. A quantitative analysis of prenatal methyl mercury exposure and cognitive development. *Am. J. Prev. Med. 29*, 353-365.

Cohen, J.T., Bellinger, D.C., Connor, W.E., and Shaywitz, B.A.. 2005b. A quantitative analysis of prenatal intake of n-3 polyunsaturated fatty acids and cognitive development. *Am. J. Prev. Med. 29,* 366-374.

Daviglus, M., Sheeshka, J.and Murkin, E. 2002. Health benefits from eating fish. *Comments Toxicol. 8,* 345-374.

Dellinger, J.A. 2007. Exposure assessment and initial intervention regarding fish consumption of tribal members of the Upper Great lakes Region in the United States. *Environ. Res. 95,* 325-340.

Food and Drug Administration (U.S. FDA). FDA consumer advisory. Available: http//www.fda.gov/bbs/topics/ANSWERS/2000/advisory.html. Washington, DC: U.S. Food and Drug Administration (accessed 1 December 2001);2001.

Food and Drug Administration (U.S. FDA). FDA consumer advisory. Available: http//www.fda.gov/bbs/topics/ANSWERS/2000/advisory.html. Washington, DC: U.S. Food and Drug Administration (accessed 1 January 2004) ;2003.

Food and Drug Administration (U.S. FDA). Mercury Levels in Commercial Fish and Shellfish. Washington, DC, Food and Drug Administration. Available http//www.fda.gov/bbs/topics/ANSWERS/2000/advisory.html. Washington, D.C.: U.S. Food and Drug Administration (accessed 1 January 2005);2005.

Food and Drug Administration (U.S. FDA). Mercury Concentrations in Fish: FDA Monitoring Program 1990-2004. Updated January 2006. Available http//www.fda.gov/~dms/admehg3.html. Washington, D.C.: U.S. Food and Drug Administration (accessed 2 December 2007), 2006.

Gochfeld, M. 2003. Cases of mercury exposure, bioavailability, and absorption.
*Ecotoxicol. Environ. Saf. 56,*174-179.

Hamrick, K. and Smith, J. Subsistence food use in Unalaska and Nikolski. Anchorage, AK: Aleutian Pribilof Island Association, 2003.

Harris, S.G. and Harper, B.L.1997. A Native American exposure scenario. *Risk Anal. 17,* 789-795.

Harris, S.G. and Harper, B.L. 2000 Using eco-cultural dependency webs in risk assessment and characterization of risks to tribal health and cultures. *Environ. Sci. Pollut. Res. 2,* 91-100.

Harper, B.L. and Harris, S.G. 2008. A possible approach for setting a mercury risk-based action level based on tribal fish ingestion rates. *Environ. Res.*in press

Hightower, J. and Moore, D. 2003. Mercury levels in high-end consumers of fish. *Environ. Health Perspect. 111,* 604-608.

IOM (Institute of Medicine) 2006. Seabood choices: balancing benefits and risks. Washington, DC. National Academy Press.

Knuth, B., Connelly, N.A., Sheehka, J.and Patterson, J. 2003. Weighing health benefits and health risk information when consuming sport-caught fish. *Risk Anal. 23,* 1185-1197

Konig, Q., Bouzan, C., Cohen, J.T., Connor,W.E., Kris-Etherton, P.M., Gray, G.M., Lawrence, R.S., Savitz, D.A. and Teutsch, S.M. 2005. A quantitative analysis of fish consumption and coronary heart disease mortality. *Am. J, Prev. Med. 29,* 335-346.

Lange, T.R., Royals, H.E. and Connor, L.L. 1994. Mercury accumulation in largemouth bass (Micropterus salmoides) in a Florida Lake. *Arch. Environ. Contam. Toxicol. 27,* 466-471.

McMichael, A.J. and Butler, C.D. 2005. Fish, health, and sustainability. *Am. J. Prev. Med. 29,* 322-323.

National Research Council (NRC). 2000. Toxicological effects of methylmercury. Washington DC, National Academy Press.

NYSDOH (New York State Department of Health). 2005. Hudson Basin Biota Database on Contaminants. New York State Department of Conservation, Albany, New York.

Patterson, J. 2002. Introduction---comparative dietary risk: balance the risks and benefits of fish consumption. *Comments Toxicol. 8*, 337-344.

Ramos, A.M.and Crain, E.F. 2001. Potential health risks of recreational fishing in New York City. *Ambulatory Ped.1*, 252-255.

Rice, G., Swartout, J., Mahaffey, K.and Schoeny, R. 2000. Derivation of U.S. EPA's oral reference dose (RfD) for methylmercury. *Drug Chem. Toxicol.* 23,41-54.

SAS (Statistical Analysis System.1995. SAS users' guide SAS Institute, Cary, North Carolina.

Silver, E., Kaslow, J., Lee, D., Lee, S., Tan, M.L., Weis, E. and Ujihara, A. 2007. Fish consumption and advisory awareness among low-income women in California's Sacramento-San Joaquin Delta. *Environ. Res. 104,* 410-419.

Stumpff, L.M. 2006. Reweaving earth: an indigenous perspective on restoration planning and the National Environmental Policy Act. Environ. *Pract. 8,* 93-103.

Toth, J.F. Jr. and Brown, R.B. 1997. Racial and gender meanings of why people participate in recreational fishing. *Leisure Sci. 19,* 129-146.

Willett, W.C. 2005. Balancing health risks and benefits. *Am. J. Prevent. Med.29,* 321-321.

In: Fish Consumption and Health
Editors: George P. Gagne and Richard H. Medrano
ISBN 978-1-60741-151-2

Chapter IV

Fish Consumption: A Paradox of Good Health

*Johnny O. Ogunji**
Department of Fisheries and Aquaculture, Ebonyi State University
P.M.B 053 Abakaliki Nigeria

Abstract

Fish is an important part of a healthy diet. There are many nutritional benefits from including fish in the diet. Fish are low in saturated fats and contain high quality protein and omega-3 fatty acids (a class of fatty-acids that are essential for cell development and have other beneficial health effects). Fish also supply a number of vitamins and minerals including calcium, iron, zinc, vitamin A, niacin, vitamin B6 and vitamin D. Eating fish can contribute to a healthy heart and a child's proper growth and intellectual development. Studies have shown that eating fish regularly (such as twice per week) can reduce the chance of death from a heart attack. However, concerns have long existed about whether fish are safe to eat due to the trace amounts of chemicals that may be present in fish tissue. Low levels of natural and man-made chemicals are in the water in which fish live. Three of the chemicals that can accumulate in fish tissue are mercury, polychlorinated biphenyls (PCBs), and chlordane among others. Human health studies have indicated that reproductive function may be disrupted by exposure to these chemicals; neurobehavioral and developmental deficits occur in newborns and continue through school-aged children who had in utero been exposed. Other systemic effects (e.g., self-reported liver disease and diabetes, and effects on the thyroid and immune systems) are associated with elevated serum levels; and increased cancer risks are also associated with exposures. This article examines the economic importance of fish consumption in relation to good health and looks at the potentials of producing safe fish through aquaculture.

---

* Correspondence: Tel. 002348067558863 Email Ogunjijo@yahoo.com.

1. World Fish Production and Consumption

According to FAO (2002) global production from capture fisheries, aquaculture and the food fish supply is currently the highest on record and remains very significant for global food security, providing more than 15 percent of total animal protein supplies. Global capture of fish for food jumped from 44 million tonnes in 1973 to 65 million tonnes in 1997 (Delgago et al., 2003). The global production in 2005 was 141.4 million tonnes. Capture fisheries contributed 95.6 million tonnes in 2000 but decreased to 93.3 million tonnes around 2005. Aquaculture on the other hand increased from 35.5 million tonnes in 2000 to 48.1 million tonnes in 2005 (FAO 2006a, b).

Fish and fisheries market has become one of the world's fastest growing international commodity markets. The global trade value has grown from USD 6 billion in 1980 to USD 56 billion in 2001 (Scholz 2005). More than 200 countries supply the global market with more than 800 commercially relevant species of fish crustacean, mollusc and squid. The product variety ranges from canned tuna to boneless salmon fillet, salted herring, dried cod, frozen fish fingers, smoked mackerel, and life lobsters, to oysters in sauce or sardines for processing into animal feed.

In 1970 the apparent world average food fish per caput consumption was estimated at 10.9 kg/year. It increased to 13.6 in 1989 and declined to 13.0 Kg/year in 1993 (Josupeit 1994). Based on the information available in August 2006, FAO has calculated the apparent world average food fish per caput consumption in 2007 as 16.4 Kg/year (FAO 2007). It can be seen that the change in growth has not been continues but that changes have existed. Interestingly current data has been the highest for quite a long time.

The per-caput fish supply is currently however, not equally distributed by economic grouping like in the past. Developed countries record a per-caput fish supply of 23.7 Kg/year, Industrialised countries – 29.5 Kg/year; developing countries – 14.5 Kg/year and low-income food – deficit countries – 13.9 Kg/year. The value for developing countries has increased from 6.2 Kg/year in 1970 to 9.4 Kg/year in 1990 (Josupeit 1994). This progressive increase is encouraging but calls for concerted effort to produce and supply safe fish for good health.

Western Europe is the main fish consuming region among developed countries with a fish per – caput supply 30.5 Kg/year. Faeroe Island and Iceland situated in this region recorded the highest per-caput fish supply as follows: Faeroe Island 87.5 and Iceland 91.4 Kg/year. Northern America 22.7 Kg/year and Oceania (Australia and New Zealand) 22.9 Kg/year recorded lower supply. Maldives has over the years become a country with highest per-caput fish supply in the world. It was 130 kg/year in 1990 and 190Kg/year currently (FAO 2007). Practically fish is the only food that does not have to be imported and it is a staple food in this country.

The foregoing has shown the increased interest of humans to consume fish. The issue to be dealt with is the safety of the fish consumed and how to produce safe fish. Before dwelling on that it is good to assess the benefits derived by man from consuming fish.

2. Benefits of Fish Consumption

Research has revealed that fish (finfish or shellfish) is a healthy food. This very important attribute of fish is linked to the type of fatty acid fish contains. Fatty acids are components of lipids. For clarity, better understanding and appreciation of the facts it is good to briefly look at how the fatty acids are categorised.

According to New (1987) fatty acids are given a common name but are also, besides their straightforward chemical formula, given a specific numerical designation, such as 14:0; 20:1; 18:3n-3; 18:2n-6; 20:4n-6; 22:5n-6; or 22:6n-3. This nomenclature refers to the length of the carbon chain in the molecule, the number of carbon-carbon double bonds present and the position of the first double bond. This can be illustrated by the chemical formulae of some of the fatty acids mentioned above. The methyl group is shown in the formulae below as CH_3.

In the designation 20: n-6, for example, '20' means that there are 20 carbon atoms in the chain. '4' means that there are four carbon-carbon double bonds (the double bond is shown as a = sign and the carbon atom is shown as C). n-6 means that the first double bond, numbering from the methyl (CH_3) end, occurs after the sixth carbon atom in the chain. Myristic acid (14:0) has fourteen carbon atoms but no C=C double bonds. Arachidic acid (20:0) also has no double bonds. Linolenic acid (18:3n-3) has eighteen carbon atoms and three double bonds, the first of which appears on the third carbon atom. Linoleic acid (18:2n-6) has two double bonds and the first occurs on the sixth carbon atom and so on.

Those fatty acids which have their first double bond on the third carbon atom are known as the 'n-3' series or the 'linolenic' series after the name of the fatty acid in the series with 18 carbon atoms in its chain. Similarly, those which have their first double bond on the sixth carbon atom are known as the 'n-6' series or the 'linoleic series.

Linolenic acid: 18:3n-3

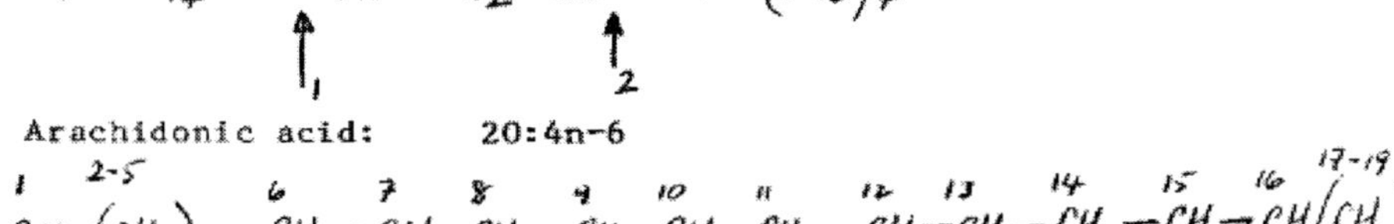

Linoleic acid: 18:2n-6

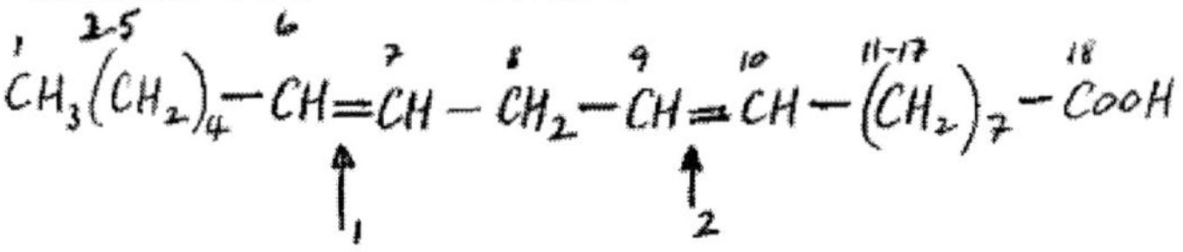

Arachidonic acid: 20:4n-6

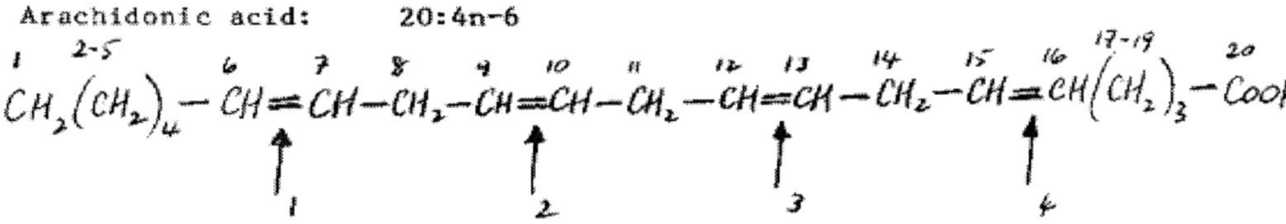

Saturated fatty acids are those without any double bonds. Monosaturated fatty acids are those with only one double bond, while those with more than one double bond are known as poly-unsaturated fatty acids. Those with fewer double bonds are referred to as 'more saturated' than those with a greater number. The n-3 series and n-6 series fatty acids, and the n-7 and n-9 fatty acids are all members of the group known as polyunsaturated fatty acids, because they have more than one double bond. Sometimes these are referred to as PUFA's for short. Members of this group which have many (4 or more) double bonds are sometimes referred to as higher unsaturated fatty acids (HUFA'S).

Evidence from several research works, including animal – experimental, observational, and clinical studies have further confirmed the importance of fish consumption in human nutrition. Two long chain n-3 polyunsaturated fatty acids (n-3 PUFAs), eicosapentaenoic acid (EPA) and docosahexaenoic acid (DHA), have been identified as the likely active constituents.

The role of n-3 fatty acids in growth and development as well as in health and disease—particularly coronary heart disease (CHD) is currently one of the fastest growing research areas in nutritional science. As a result, knowledge of these essential PUFAs has grown significantly (Simopoloulos 1999). Benefits from n-3 fatty acids were first noticed in epidemiological studies among Greenland Inuit. The rarity of ischemic heart disease (characterized by narrowed heart arteries) in this Eskimo population was attributed to their traditional, EPA-rich diet of marine animals and fish. Moreover, Eskimo consumption of n – 3 fatty acids was low compared to the typical Danish diet (Bang et al., 1980). This observation has led to several epidemiological and observational studies of men and women, who regularly consume fish, showing similar heart health results (Kris-Etherton 2002). Cardiovascular benefits from n-3 fatty acid consumption have been documented in several prospective studies and randomized clinical trials (Harris and Isley, 2001; Kris-Etherton et al., 2002; Von Schacky et al., 2002; AHRQ 2004).

Von Schacky et al., 2002 in a clinical trial found that patients who were given fish oil concentrate for two years had lower triglyceride levels and minimal artery blockage when compared to those receiving a placebo. These results showed that consuming n-3 fatty acids daily may reduce the risk of abnormal thickening of arteries due to fatty deposits on arterial inner walls (atherosclerosis). n-3 PUFAs influence several cardio vascular risk factors (Mozaffarian, 2006). Effects occur within weeks of intake and may result from altered membrane fluidity and receptor responses following incorporation of n-3 PUFAs into cell membranes (Clandinin et al., 1991; Feller and Gawrisch 2005) and direct binding of n-3 PUFAs to intracellular receptors regulating gene transcription (Vanden 2004).

The heterogeneity of the effects of fish or fish oil intake on cardiovascular outcomes is likely related to varying dose and time responses of effects on the risk factors. At typical dietary intakes, antiarrhythmic effects predominate, reducing risk of sudden death and CHD death within weeks. At higher doses, maximum antiarrhythmic effects have been achieved, but other physiologic effects may modestly impact other clinical outcomes (possibly requiring years to produce clinical benefits). For instance, nonfatal myocardial infarction may not be significantly affected by lower doses or shorter durations of intake but may be modestly reduced by higher doses or prolonged intake (e.g., 1.8 g/d for 5 years17) (Mozaffarian and Rimm 2006).

A meta-analysis of studies on diabetes and fish oils found a 30 percent reduction in patient triglyceride levels—particularly among subjects with Type I diabetes. This is noteworthy, considering that the high cardiovascular mortality rate of people with diabetes has been partially attributed to increased triglyceride concentrations in the blood (Friedberg 1998). It has been suggested that n-3 fatty acids have a protective mechanism of reducing the risk of irregular heartbeats (De Lorgeril et al., 1999; Marchioli et al., 2002; Singh et al., 2002). Marchioli et al. (2002) demonstrated that even a small amount (1 g/d) of n-3 fatty acids is effective in reducing overall mortality and risk of cardiac death by 20 percent and 45 percent, respectively.

Omega-3 fatty acids are also shown to be beneficial at various life stages and with several health conditions. DHA is naturally found in breast milk and has been shown to support the visual and cognitive development in infants. Various infant formulas now contain DHA along with arachidonic acid (20:4n-6; got from eating meat, poultry, and eggs) to more closely mimic breast milk. Evidence has also emerged showing benefits from n-3 fatty acid consumption in health conditions such as, depression, cancer, lupus, and asthma. A cross-sectional study in Australia showed an inverse association between age-related maculopathy (ARM) — a leading cause of blindness in older adults — and fish consumption. Individuals who ate fish more than once per week reduced their risk of having ARM by half compared to those who ate fish less than once per month. Patients with rheumatoid arthritis, who received fish oil supplements in addition to background medication, had a significant decrease in the number of tender joints and morning stiffness (Reifen et al., 1998; Kremer 2000; Smith et al., 2000; Koumenis et al., 2003; Jho et al., 2004; Suzuki et al., 2004). The continued study of n-3 fatty acids will lead to further understanding of their effects on development, health, and disease risk reduction.

Fish also supply a number of vitamins and minerals including calcium, iron, zinc, vitamin A, niacin, vitamin B6 and vitamin D. Eating fish can contribute to a child's proper growth and intellectual development (IDPH 2005). It lowers the risk of some cardiovascular outcomes (Mozaffarian and Rimm 2006), dementia, (Morris et al., 2005) neuropsychiatric disorders (Peet and Stokes, 2005; Young and Conquer 2005) and inflammatory disorders (Mori and Beilin, 2004; Mickleborough et al., 2006).

3. Fish Contamination by Chemicals and Effect of Consumption

There have been concerns whether fish are safe to be eaten due to the trace amounts of chemicals that may be present in fish tissue. Low levels of natural and man-made chemicals are in the water in which fish live. Three major chemicals that can accumulate in fish tissue to produce deleterious effect on human are mercury, polychlorinated biphenyls (PCBs), and chlordane. It is important to trace how these contaminants get into the water naturally and through pollution.

The Joint Group of Experts on Scientific Aspects of Marine Pollution (GESAMP) (1980) defined marine pollution as "the introduction by man, directly or indirectly of substances or energy (e.g. heat) into the marine environment (including estuaries) resulting in such deleterious effects as harm to living resources, hazards to human health, hindrance to marine activities including fisheries, impairment of quality for use of seawater and reduction of amenities". This definition, although given with special consideration to marine water will be applied here generally to all water or aquatic environment (marine and fresh water).

Very important pollutants are heavy metals, that include mercury (Hg), cadmium (Cd), lead (Pb), chromium (Cr), copper (Cu) and zinc (Zn). Their natural sources are by water runoffs from areas with rocks containing these heavy metals. The man made sources are from the production plants of these metals as dusts and water runoffs; the dumping of refuse from factories and workshops which use these chemicals; from domestic waste through improper disposal of batteries and other such materials; weathering of Zinc roofs; Cu and Cd plating from paper producing factories and agriculture, these metals are also found in fungal agents for pulp and seed. They are also found in sewage sludge. Pb is also generated from vehicle exhaust for it is a petrol additive.

Heavy metals tend to be absorbed to mud particles and thus vanish into the bottom of lakes, rivers and sea. As long as these "sinks" are not disturbed the heavy metals are then no longer bio-available. However, these pollutants are taken up directly from the water and they accumulate in food chain. Among solutes present in surface and ground water, heavy metals in their ionic forms belong to a category of pollutants. Their accumulation in the tissues of an organism can result in chronic illness. Being non-biodegradable, they can be concentrated along the food chain, producing their toxic effects at points often far removed from the source of pollution (Alam et. al., 1983).

Halogenated hydrocarbons are another major group of water pollutants. These are organic compounds, where hydrogen (H) has been (partly or fully) substituted by chlorine and/or fluorine. Their major sources are pesticides, technical fluids like hydraulic "oil" and brake fluid, cooling "oil" and softener in plastic materials, some plastic materials (PVC) and dry cleaning fluid. Pesticides can be classified on the basis of their intended pest targets. These categories are given by Entomological Society of America (ESA) (1980), as follows:-

- Fungicides, used to protect crop plants and animals from fungal pathogens e.g. organometalic compounds of mercury and tin chlorophenolics as tri- tetra- and penta- chlorophenol.
- Herbicides, used to kill weedy plants, so as to release desired crop plant competition chloroaliphatics such as dalapon and trichloroacetate.
- Insecticides, used to kill insects pest and vectors of deadly animal diseases such as malaria, yellow fever, tripanosomiasis (e.g. the DDT group of chlororinated hydrocarbons including DDT, DDDand methoxychlor).
- Acaricides, used to kill mites, which are pests in agriculture and ticks that carry encephalitis of human and domestic animals.
- Molluscides are used against snails and slugs that can be important pests of citrus groves and vegetables and flower gardens. They are also used to kill aquatic snails which are vectors of certain human diseases especially schistosomiasis. Examples are ethylene dibromide, dichloropane or dichloropropane.
- Others are rodenticides used for the control of rats, avicides, for the control of birds and antibiotics used for control of bacteria caused infections band diseases.

Halogenated hydrocarbons are found in water only in low concentrations. They attach themselves readily to organic matter, so they are kept within the flood chains for a long time until they are eventually degraded. These pollutants are lipophhilic, so they easily get dissolved in fat tissue and accumulate in food chain.

Crude and mineral oil are petroleum products which have become very outstanding water pollutants. Crude oil is compounds that occur in liquid form. They are obtained by drilling. Through the process of refining, various hydrocarbon fractions are separated by fractional distillation. The refined hydrocarbons are referred to as minerals here, and include fuels (diesel, kerosene, bunker oil) and all lubricants (engine oil etc). The sources of crude oil pollutants into the aquatic environment include; during on-site drilling and production, during storage, shipment and processing through spillage or leakage. The source of the mineral oils are from processing plants, fuel from tanks are some time washed into water. Other times it is during shipment and from all sorts of powered vehicles. Bunker is a very heavy fuel used to power big ships, it contains plenty of unusable substances, when it is purified on board the vessels, the residual is frequently dumped over board. These more or less are the main sources of the present day "tar balls" in the sea. Lubricants dripping from vehicles and other engine installations, including used engine oil that are dumped indiscriminately are also major sources. Only a small fraction of these pollutants dissolve in the water especially the non-aliphatics, which are poisonous.

From towns and crude oil production sites oil is flushed into the water with rain runoff. In a case of oil spill, the oil can cover the water surface. It later disintegrates into tar balls; part of it may sink to the bottom and mix there with sediments. At beaches the waves can mix oil and water to sticky emulsion referred to as, "chocolate mouse". Some of the oil will be deposited on the mud flats.

Mercury is a naturally occurring element in the environment and is also released into the air through industrial pollution. Mercury that falls from the air can accumulate in streams, rivers, lakes and the ocean. Mercury is often the contaminant of greatest concern in fish tissue world over. It is a reactive heavy metal. In water it is converted by microbial activity into organic methylmercury. Inorganic mercury is poorly absorbed following ingestion, and elemental mercury does not readily cross tissue barriers. In contrast, methylmercury is readily absorbed and actively transported into tissues (EPA 2006). Thus, methylmercury bioaccumulates in aquatic food chains and has greater potential toxicity than inorganic Mercury (EPA 2005; 2006; RAIS 2006). Concentrations of methylmercury in aquatic species depend on levels of environmental contamination and on the predatory nature and lifespan of the species. Larger, longer-living predators (eg, swordfish, shark) have higher tissue concentrations, while smaller or shorter-lived species (eg, shellfish, salmon) have very low concentrations. Preparation methods have little impact on methylmercury content (Mozaffarian and Rimm 2006).

Health effects of very high mercury exposure following occupational or industrial accidents are well documented. These include paresthesias, ataxia, and sensory abnormalities in adults, and delayed cognitive and neuromuscular development following in utero exposure (Gochfeld, 2003). Toxicity appears related to binding of methylmercury to sulfhydryl groups of enzymes, ion channels, and receptors, resulting in inhibition of antioxidant systems and production of free radicals and reactive oxygen species (NRC 2000).

Neurotoxicity is generally considered the health impact of greatest concern for dietary methylmercury intake. High-dose human exposure can result in sensory and motor impairment and other adverse impacts, up to and including death. High dose in utero exposure can cause mental retardation, cerebral palsy and deafness. There is also some evidence suggesting lower-dose exposures can have cardiovascular and immunological effects, but good epidemiological data is lacking. There is no argument that high doses of methylmercury can have severe health impacts. The question is whether relatively low level exposure, such as that incurred by fish consumption, presents more subtle risks such as cognitive disabilities, especially for children exposed to low levels in utero and though early childhood consumption of mother's milk and fish. The Iowa Department of Public Health (IDPH), using USEPA guidance, determined that all fish containing less than 0.3 ppm mercury are safe to consume with no meal restriction. Consumption of up to 1 meal/week of fish containing between 0.3 and 1.0 ppm of mercury was safe and fish containing over 1.0 ppm of mercury should not be eaten (which is consistent with the current mercury FDA action level) (IDPH, 2005).

Chlordane is an organochlorine pesticide that was produced and sold in the United States from 1946 through 1987. Chlordane is resistant to breakdown in the environment and has a half-life of 10-20 years. Due to known toxicity and potential carcinogenicity, the United States Environmental Protection Agency (USEPA) cancelled registration for all uses of chlordane on food crops in 1978. Before 1978, chlordane was used extensively as a lawn and garden pesticide and for the control of cutworms in corn. After restrictions were placed on usage in 1978, the primary use for chlordane was as a termaticide in building construction. Chlordane has been shown to cause liver cancer in mice and rats. USEPA classifies chlordane as a "probable" human carcinogen. In 1988, due to continued controversy over its use, the USEPA cancelled the registration of all products containing chlordane. Use of existing supplies was allowed until April 15, 1988 when both the sale and use of technical chlordane were voluntarily suspended. The sale, purchase, and use of chlordane in Iowa were banned as of January 1, 1989. The IDPH, using USEPA guidance, determined that all fish containing less than 0.6 ppm chlordane are safe to consume with no meal restriction. Consumption of up to 1 meal/week of fish containing between 0.6 and 5.0 ppm of chlordane was safe and fish containing over 5.0 ppm of chlordane should not be eaten.

Polychlorinated biphenyls (PCBs) are industrial chemicals that were manufactured in the United States from 1929 until 1977 under the trade name Aroclor. Currently, PCBs are not produced in the United States, but many products containing PCBs are still in service. Dioxins— commonly referring to dibenzodioxins and dibenzofurans—are organochlorine by-products of waste incineration, paper bleaching, pesticide production, and production of polyvinyl chloride plastics. Prior to 1977, PCBs were used in closed electrical systems such as capacitors and transformers. Before 1971, PCBs were used in a number of applications, including plasticizers, heat transfer fluids, hydraulic fluids, fluids in vacuum pumps and compressors, lubricants and wax extenders. Similar to chlordane, PCBs are also organochlorine compounds that are resistant to breakdown in the environment. PCBs tend to adsorb to sediments or other organic matter in aquatic environments. PCBs are highly lipophilic (fat soluble) and accumulate in tissues of fish and other aquatic organisms by a factor of 10,000 over concentrations found in ambient waters. PCBs are classified by USEPA as "probable" human carcinogens. The IDPH, using USEPA guidance, determined that all fish containing less than 0.2ppm PCBs are safe to consume with no meal restriction. Consumption of up to 1 meal/week of fish containing between 0.2 and 2.0ppm of PCBs was safe and fish containing over 2.0ppm of PCBs should not be eaten (which is consistent with the current PCBs FDA action level) IDPH (2005).

Animal experiments and some evidence in humans indicate that PCBs and dioxins are carcinogenic, possibly related to effects on the aryl hydrocarbon receptor, a transcription factor affecting gene expression (WHO 1998). According to Mozaffarian and Rimm (2006) multiple congeners (structural variants) of PCBs and dioxins exist. Potential toxicities of foods are calculated using toxic equivalence (TEQ): the sum of each congener's level in the food multiplied by that congener's toxic equivalency factor (standardized against 2,3,7,8-tetrachlorodibenzo-p-dioxin). In the United States, PCBs comprise 28% and dioxins 72% of total TEQ exposure (Schecter et al., 2001). Among adults, major dietary sources of PCBs and dioxins are beef, chicken, and pork (34% of total TEQ); dairy products (30%); vegetables (22%); fish and shellfish (9%); and eggs (5%). However, a considerable attention has been given to fish sources. Dietary sources are similar for children. When PCBs and dioxins were measured in farmed and wild salmon (Hites et al., 2004; Hamilton 2005) the levels were similar to those in several other foods (WHO 1998).

Prenatal exposure to PCBs and dioxins has been associated with childhood neurodevelopmental deficits in several (Jacobson and Jacobson 1996; Patandin et al., 1999; Ribas-Fito et al., 2001; Stewart et al., 2003; Schantz et al., 2003) but not in all the studies (Buck et al., 2003). Because most exposure (> 90%) generally comes from meat, dairy, and vegetable sources this concern is not specific to fish consumption, particularly since fish also contains potentially beneficial DHA. However, women consuming one or more servings per day of commercial freshwater fish or consuming locally caught freshwater fish from highly contaminated inland sources may be more greatly exposed to PCBs and dioxins (Judd et al., 2004). It has been reported that the PCB content in fish can be reduced 12% to 40% by trimming belly and back fat during filleting and by not consuming the skin (Thannum 2006). High contaminant levels are typically measured in unprepared foods, and cooking may reduce PCB and dioxin content (Schecter et al., 1998).

4. Aquaculture and Aquaculture Production

Aquaculture is the rearing of aquatic organisms under controlled or semi-controlled conditions. Thus aquaculture is underwater agriculture. According to the definition used by FAO for statistical purposes, aquaculture is the farming of aquatic organisms including fish, molluscs, crustaceans and aquatic plants. Farming implies some sort of intervention in the rearing process to enhance production, such as regular stocking, feeding, protection from predators, etc. Farming also implies individual or corporate ownership of the stock being cultivated

Aquaculture is currently one of the fastest growing food production systems in the world. With stagnating yields from many capture fisheries and increasing demand for fish and fishery products, expectations for aquaculture to increase its contribution to the world's production of aquatic food are very high, and there is also hope that aquaculture will continue to strengthen its role in contributing to food security and poverty alleviation in many developing countries. However, it is also recognized that aquaculture encompasses a very wide range of different aquatic farming practices with regards to species (including seaweeds, molluscs, crustaceans, fish and other aquatic species groups), environments and systems utilized. The systems have very distinct resource use patterns involved. They offer a wide range of options for diversification of avenues for enhanced food production and income generation in many rural and peri-urban areas.

In view of the significant health cum nutritional, social, economic and environmental benefits, which generally can be associated with most existing aquaculture practices, and the good prospects for further development and expansion of the sector, it is important for efforts aiming at the sustainable development of aquaculture that potential social conflicts and environmental problems are minimized.

A striking difference between aquaculture and more traditional aspects of fishery management according to Stickney (1979) is associated with degree of control exerted by man over the environment. Fishery managers attempt to manage natural aquatic ecosystems, usually for the benefit of sport or commercial fishermen. In general, the aquaculturist exercises much more control on the environment than the fishery manager. The aquaculturist manages water systems for maximum production of one or small number of species and attempts to eliminate insofar as possible all competition and sources of mortality.

Aquaculture's contribution to global supplies of fish, crustaceans and molluscs continues to grow increasing from 3.9 percent of total production by weight in 1970 to 27.3 percent in 2000 (FAO 2002) and 51.63 percent in 2005 (FAO 2006b). Absolute growth of aquaculture is more rapid than all other animal food producing sectors.

During the past three decades, aquaculture has expanded, diversified, intensified and made technological advances. The potential of this development to enhance local food security, alleviate poverty and improve rural livelihoods has been well recognized. The need for the aquaculture sector to continue development towards its full potential, and making a net contribution to global food availability, domestic food security, economic growth, trade and improved living standards has been emphasised (FAO 2002).

The ability of producing safer fish from aquaculture should become the next target in world fish production. This is possible because the water environment through which fish becomes tinted with chemicals and pollutants causing health concerns to man can be screened before use in aquaculture for fish production.

5. Safe Fish Production in Controlled Environment

FAO (1997) defines the term sustainable development as the management and conservation of the natural resource base and the orientation of technological and institutional change in such a manner as to ensure the attainment and continued satisfaction of human needs for present and future generations. Such sustainable development (in the agriculture, forestry and fisheries sectors) conserves land, water, plant and animal genetic resources, is environmentally non-degrading, technically appropriate, economically viable and socially acceptable.

"Attainment and continued satisfaction of human needs" in fish production calls for producing fishes that may not predispose anyone to disease or ill health. Such fish is here referred to as "safe fish". A situation where the sum total of fish contamination by deleterious chemicals cannot be completely isolated from the contamination in water the need to consciously use contamination free water in fish production has become paramount.

Though the world has become aware of the intricacies of water pollution and contamination by heavy metals cum organic compounds, not so much has changed. The consciousness to produce safe fish should become ingrained in the mind of fish producers and in government policies. Aquaculture due to its peculiarity stands out as the only production system through which safe fish can be produced. There is the possibility of screening the production water source even before use. In this way contaminated water sources are not utilised but discarded. Policies should be formulated through world fisheries regulating bodies and governments to include safe fish production as a good aquaculture practice that can be enforced.

The concept of safe fish production should be publicised and enforced. It can also be integrated into the Code of Conduct for Responsible Fisheries (CCRF). This code was development by FAO in consultation and collaboration with relevant United Nations Agencies and other international organizations, including non-governmental organizations. It was adopted by 31st October 2005. Article 9.1 of the code, deals with responsible development of aquaculture, including culture-based fisheries, in areas under national jurisdiction (FAO 1997). Safe fish production can become an integral part of responsible aquaculture practices.

States were given the right under the code to designate or establish an authority or authorities competent, empowered and capable to effectively promote support and regulate aquaculture and culture-based fisheries. Appropriate institutional linkages with other authorities such as those concerned with agriculture, rural development, water resources, environment, health, education and training and many others, should also be established. These linkages may have to be expressed in legislative forms.

In addition to appropriate regulation and control fish producers should understand the risk in consuming contaminated fish and thus decided to produce safe fish. The benefits will be enjoyed both by the fish producers and consumers.

On the other hand world oceans, seas and rivers should be maintained for the benefit of man. Interestingly fisheries governance has been affected by the deliberations of the United Nations Conference on the Environment and Development (UNCED), held in 1992 in Rio de Janeiro, Brazil. Since that conference, the ecosystems dimensions of management issues have received increasing attention from governments, the UN system and the fishing industry. All are now more ready to recognize that fish are an integrated part of an aquatic ecosystem, a system in which modifications in one area have the potential to affect other areas. Thus, it is increasingly regarded as necessary, first to monitor the state of the aquatic ecosystem, and then to manage human interventions within that ecosystem. Only within such a framework will it be possible for capture fisheries to continue to be a source of food and income for future generations. In the coming years, humankind is likely to improve dramatically its understanding of the intricacies of aquatic ecosystems, and this will lead not only to more knowledge but also, and perhaps paradoxically, to growing uncertainties. As a result, there will be growing pressure for a strict application of the precautionary approach to all interventions, including those by fishers, in aquatic ecosystems. Both the aquaculture industry and the capture fisheries sector will realize that they must be seen by all concerned to respect this principle.

Conclusion

Continued efforts should be sustained to limit environmental contamination from heavy metals and organochlorine compounds. Attempts should be made to produce safe fish (fishes produced under aquaculture environment with uncontaminated water). It has been stated that the potential risks of fish intake must be considered in the context of possible benefits. Based on strength of scientific research evidence (though a bit conflicting) and probable degree of effect, the benefits of careful fish consumption surpass the risks among adults. A few fish species can be too risky for women of childbearing age to consume. This is dependent on the ecology of the fish species and the pollution level of the water bodies. Information about the contamination status of fish species in each locality should be sought. However, avoidance of modest fish consumption due to confusion regarding risks and benefits could result in thousands of death annually due to coronary heart disease and suboptimal neurodevelopment in children.

Acknowledgment

My gratitude goes to Dr Mozaffarian, D. who sent to me on request his article that was very informative. I am grateful to Mr Ede Hyacinth for proof reading the manuscript and making necessary corrections. Other colleagues who made relevant suggestions are thanked. I am highly indebted to Mrs Victoria Ogunji my wife and children Victory, Nicholas and Sylvia.

References

AHRQ (Agency for Healthcare Research and Quality) (2004). US Department of Health and Human Services. March 2004. AHRQ Pub. No. 04-E009-1.

Alam, A. M. S., Haque, A. N. M., Ali, M. Y., Tarafdar, S. A., Khan, A. H., (1993) Heavy elements contaminants in aquatic ecosystems. In: Hypertrophic and polluted freshwater ecosystem: Ecological bases for water resource management. *Proc. Int. Symp. Limnol.* 25 – 28 Nov. Dept. of Botany. University of Dhaka, Bangladesh.

Bang H. O., Dyerberg J, Sinclair H. M. (1980) The composition of the Eskimo food in northwestern Greenland. Am. J. Clin. Nutr.; 33: 2657 – 2661.

Clandinin M. T., Cheema S, Field C. J., Garg M. L., Venkatraman J, Clandinin T. R. (1991) Dietary fat: exogenous determination of membrane structure and cell function. *FASEB J.* 5: 2761 – 2769.

NRC (2000). Toxicological Effects of Methylmercury. Committee on the Toxicological Effects of Methylmercury, Board on Environmental Studies and Toxicology; Commission on Life Sciences, National Research Council.. Washington, DC: National Academies Press

Delgago C. L. Wada, N., Rosegrant M. W., Meijer, S., Ahmed M. (2003) The future of fish: Issues and trends to 2020. International Food Policy Research Institute (IFPRI), World Fish Centre, 1 – 6 pp

De Lorgeril M., Salen, P., Martin, J. L. Monjaud, I., Delaye, J., Mamelle, N. (1999). Mediterranean diet, traditional risk factors, and the rate of cardiovascular complications after myocardial infarction. *Circulation.*99: 779 – 785.

Entomological society of America (1980). Pesticide Hand book. ESA College Park M.D. FAO (1997). Technical Guidelines for Responsible Fisheries. No. 5. Rome, FAO. 1997. 40p.

FAO (2002) The state of world fisheries and aquaculture http://www.fao/docrep /003/x8002E/x8002e.HTM

FAO (2006a) Food and Agriculture organisation - disposition of world fishery production statistics. Yearbooks of Fishery Statistics Summary tables. http://www.fao.org /fishery /statistics / global- commodities - production, ftp://ftp. fao.org /fi/STAT /summary/ default.htm#aqua

FAO (2006b) Food and Agriculture organisation – Statistics of World aquaculture production by inland and marine waters Statistics. Yearbooks of Fishery Statistics Summary tables. http://www.fao.org/fishery/statistics/global-commodities-production, ftp://ftp.fao.org/fi/STAT/summary/default.htm#aqua

FAO (2007) Laurenti, G. (comp.) 1961-2003 Fish and fishery products: world apparent consumption statistics based on food balance sheets. FAO Fisheries Circular No.i821, rev.8. Rome, 429 p.

Feller S. E., Gawrisch K. (2005). Properties of docosahexaenoic-acid-containing lipids and their influence on the function of rhodopsin. *Curr Opin Struct Biol.*15:416-422.

Friedberg C. E., Janssen M. J., Heine R. J., Grobbee D. E.. (1998). Fish oil and glycemic control in diabetes. *Diabetes Care.* 21(4): 494 – 500.

GESAMP (Joint Group of Experts on the scientific Aspects of Marine Pollution) (1980) Report of the eleventh Session Report and Studies No. 10. UNDP (GESAMP – IMO /FAO /UNESCO /WINO /WHO /IAEA /UN / UNEP.

Gochfeld M. (2003) Cases of mercury exposure, bioavailability, and absorption. Ecotoxicol Environ Saf. 56: 174 – 179.

Hamilton M.C., Hites, R. A., Schwager S. J., Foran J. A., Knuth B. A., Carpenter D. O. (2005) Lipid composition and contaminants in farmed and wild salmon. *Environ. Sci. Technol.* 39: 8622 – 8629.

Harris W. S., Isley W. L. (2001). Clinical trial evidence for the cardioprotective effects of omega-3 acids. Curr Atheroscler. Rep. 3(2): 174 – 179.

Hites R. A., Foran J. A., Carpenter D. O., Hamilton M. C., Knuth B. A., Schwager S. J. (2004) Global assessment of organic contaminants in farmed salmon. *Science.* 303: 226 – 229.

IDPH (Iowa Department of Public Health in cooperation with Iowa Department of Natural Resources (IDNR) (2005). New protocol for issuing fish consumption advisories http://www.iowadnr.com/fish/news/consump.html

Jho DH, Cole SM, Lee EM, Espat NJ. (2004) Role of Omega-3 Fatty Acid Supplementation in Inflammation and Malignancy. *Integr. Cancer Ther.* 3(2): 98 – 111.

Judd N, Griffith W. C., Faustman E. M. (2004) Contribution of PCB exposure from fish consumption to total dioxin-like dietary exposure. *Regul. Toxicol. Pharmacol.* 40:125 – 135.

Josupeit H. (1994) Global overview of fish consumption. FAO/Globefish Fisheries Industry Division Rome Italy.

Kremer J. M. (2000) n-3 fatty acid supplements in rheumatoid arthritis. *Am. J. Clin. Nutr.* 71: 349S – 351S.

Kris-Etherton P, Harris WS, Appel LJ. (2002). Fish consumption, fish oil, omega-3 fatty acids, and cardiovascular disease. *Am. Heart Assoc .Sci. Statement.* 2747 – 2757.

Koumenis I. L., Edens, M. B., Tramposch K. M., Clayton B., Bowton, D., Chilton, F. H. (2003) Inhibition of leukotriene biosynthesis by a novel dietary fatty acid formulation in patients with atopic asthma: a randomized, placebo-controlled, parallel-group, prospective trial. *Clin. Ther. Mar.* 25(3): 972 – 979.

Marchioli R, Barzi F, Bomba E, Chieffo C, Di Gregorio D, Di Mascio R, Franzosi MG, Geraci E, Levantesi G, Maggioni AP, Mantini L, Marfisi RM, Mastrogiuseppe G, Mininni N, Nicolosi GL, Santini M, Schweiger C, Tavazzi L, Tognoni G, Tucci C, Valagussa F. (2002) Early protection against sudden death by n-3 polyunsaturated fatty acids after myocardial infarction: time-course analysis of the results of the Gruppo Italiano per lo Studio della Sopravvivenza nell'Infarto Miocardico (GISSI)-Prevenzione. *Circulation.* 2002 Apr 23;105(16):1897 – 1903.

Mickleborough T. D., Lindley M. R., Ionescu A. A., Fly A. D. (2006) Protective effect of fish oil supplementation on exercise-induced bronchoconstriction in asthma. *Chest.*129:39 – 49.

Mori T. A., Beilin L. J. (2004) Omega-3 fatty acids and inflammation. *Curr. Atheroscler. Rep.* 6: 461 – 467.

Morris M. C., Evans D. A., Tangney C. C., Bienias J. L., Wilson R. S. (2005) Fish consumption and cognitive decline with age in a large community study. *Arch. Neurol.* 62: 1849 – 1853.

Mozaffarian, D. (2006) Fish intake, contaminants, and human health: Evaluating the risks and the benefits. *J. Am. Md. Ass.* 296, (15): 1885 – 1899.

New, M. B. (1987). Feed and feeding of shrimp and fish. Aquaculture Development and Coordination Rogramme (ADCP/REP/87/26) United Nations Development Programme Food and Agriculture Organization of the United Nations, Rome. Pg 16 – 19.

Peet M., Stokes C. (2005) Omega-3 fatty acids in the treatment of psychiatric disorders. Drugs. 65:1051 – 1059.

Reifen R, Blank M, Afek A, et al. (1998) Flax reduces antibody levels. Lupus. 7(3): 192 – 197.

Risk Assessment Information System (2006). Toxicity summary for mercury. http://risk.lsd.ornl.gov/tox/profiles/mercury_f_V1.shtml. Accessed January 24,2006.

Schecter A., Dellarco M., Papke O., Olson J. (1998) A comparison of dioxins, dibenzofurans and coplanar PCBs in uncooked and broiled ground beef, catfish and bacon. *Chemosphere.* 37: 1723 – 1730.

Schecter A, Cramer P, Boggess K, Päpke, O., Olson, J., Silver, A., Schmitz, M., (2001) Intake of dioxins and related compounds from food in the U.S. population. *J. Toxicol. Environ. Health* A. 63: 1 – 18.

Scholz U. (2005) How important is fishery sector for the developing world? *Agriculture and Development* 12 (2): 38 – 40.

Singh R. B., Dubnov G, Niaz M. A., Ghosh, S., Singh, R., Rastogi, S., Manor, O., Pella, D., Berry, E. (2002) Effect of an Indo-Mediterranean diet on progression of coronary artery disease in high risk patients (Indo-Mediterranean Diet Heart Study): a randomized single-blind trial. *Lancet.*360:1455 – 1461

Simopoloulos, A. P. (1999). Essential fatty acids in heath and chronic disease. *Am. J. Clin. Nutr.* 70: 560S – 569S.

Smith W., Mitchell P., Leeder S.R. (2000) Dietary fat and fish intake and age-related maculopathy. *Arch. Ophthalmol.* 118: 401-404.

Stickney, R. R. (1979) Principles of warm water aquaculture. Johny Wiley and Sons New York. 375 pp

Suzuki S., Akechi T., Kobayashi M., Taniguchi K., Goto K., Sasaki S., Tsugane S., Nishiwaki Y., Miyaoka H., Uchitomi Y. (2004) Daily omega-3 fatty acid intake and depression in Japanese patients with newly diagnosed lung cancer. *Br. J. Cancer.* 23: 90(4): 787 – 793.

Thannum J. (2006) Great Lakes Indian Fish and Wildlife Commission. Tribally sold Lake Superior Fish easily meet FDA restrictions for chemical contaminants. MAZINA'IGANM: A Chronicle of Lake Superior Ojibwe. 23p. http: //www.glifwc.org /pub/summer00/fish_contaminants .htm.

US Environmental Protection Agency. (1997) Mercury Study report to Congress. http://www.epa.gov/mercury/report.htm.

US Geological Survey. (2001) Mercury in the environment. http://www.usgs.gov /themes /factsheet/146-00/.

Vanden H. J. P. (2004) Diet, fatty acids, and regulation of genes important for heart disease. Curr Atheroscler Rep. 6: 432 – 440.

Von Schacky C, Angerer P, Kothny W, Thiesen K, Mudra H. (1999) The effect of dietary n-3 fatty acids on coronary atherosclerosis. *Ann. Intern. Med.* 130: 554 – 562.

WHO (World Health Organization) (1998) Assessment of the health risk of dioxins: re-evaluation of the Tolerable Daily Intake (TDI). WHO Consultation; May 25-29, 1998; Geneva, Switzerland.

Young G, Conquer J. (2005) Omega-3 fatty acids and neuropsychiatric disorders. *Reprod. Nutr. Dev.* 45: 1 – 28.

In: Fish Consumption and Health
Editors: George P. Gagne and Richard H. Medrano
ISBN 978-1-60741-151-2

Chapter V

Protein-Energy Wasting, Chronic Inflammation and PUFAS: Metabolic and Cardiovascular Risk in ESRD

Gordana Perunicic-Pekovic[1], Zorica Rasic-Milutinovic[1], Sladjana Sobajic[2], Ivana Djuricic[2] and Radojka Maletic[3]
[1]Department of Nephrology and Endocrinology,
[2]Faculty of Pharmacy,
[3]Faculty of Agriculture,
University Hospital Zemun, Belgrade, Serbia

Abstract

The patients with end-stage renal disease (ESRD) include significant percentage of malnourished patients together with other risk factors: dyslipoproteinaemia, insulin resistance, increased oxidative stress, inflammation, and they all together impair endothelial function.

Nutrition, glicoregulation disorders and insulin resistance as etiologic factors participate a grate deal in premature atherosclerosis. Abnormal fatty acid patterns are reported in patients with such diseases as diabetes mellitus, inflammatory bowel disease, rheumatoid arthritis and atherosclerosis. Basic mechanism of these disorders is connected with changes in cells functions on the membrane level. These basic discoveries help us in understanding the damage process of endothelial and appearance of its dysfunction by inflammation process and by lipid peroxidation process.

We examine association between atherosclerotic risk factors and nutritional status in hemodialysis patients. Our research is a part of the study in which haemodialysis patients will be followed. In our previous study we examined HD patients which were administered supplementation n-3 PUFAs for two months. Dietary regime with fish oil could be used in dialysis patients to slow down atherosclerosis process and improve nutritional parameters.

Forty three hemodialysis patients were examined (20 males, 23 females, 55±12 years). Nutritional and inflammatory markers-including serum concentrations of C-reactive protein, tumor necrosis factor alpha, and interleukin 6- were measured. Mortality was followed prospectively for up to 12 mo.

Significant negative correlation between serum level of CRP and macroangiopathy was found. There was significant positive correlation between arachidonic acid (AA) and high sensitivity C-reactive protein (hs-CRP). Significant correlation was found between body mass index (BMI) and n-3 fatty acids.

Increasing inflammation and endothelial dysfunction predict the development of vascular disease. We report the relationship between inflammatory markers and nutritional parameters indices of atherosclerosis and other cardiovascular risk factors in patients on hemodialysis.

Keywords: malnutrition, inflammation, hemodialysis, PUFAs

Introduction

Malnutrition and inflammation are associated with end-stage renal disease (ESRD). Interleukin (IL-6) and tumor necrosis factor alpha (TNF-α) powerfully predict death from cardiovascular disease in dialysis patients as well as progression of vascular injury. Atherosclerosis and inflammation have similar basic mechanisms involving the adhesion of leukocytes to vascular endothelium in their early phases. Major predictors of clinical outcome in dialysis patients are protein-energy malnutrition (PEM) and inflammation. A common mechanism for the development of cardiovascular disease and malnutrition in dialysis patients may be cytokine activation.[1,2,3,4]

Many studies were examined inflammatory markers in process endothelial dysfunction. Inflammatory state in end stage renal disease (ESRD) are associated with an elevation of serum acute phase proteins, including CRP and some pro-inflammatory cytokines such as IL-6 and TNF alpha. IL-6 is reported to play a central role in the pathophysiology of adverse effects of inflammation in ESRD patients. Increased activation of inflammatory cytokines, such as IL-6 and TNF-alpha, may cause muscle breakdown and hypoalbuminemia, and may be involved in atherogenesis.[5,6,7,8,9,10]

Dyslipidemia is one of the possible risk factors for advanced atherosclerosis in patients with chronic renal failure. Abnormal phospholipid metabolism may play an important role in the progression of atherosclerosis in patients with renal failure. The total concentration of serum non-esterified fatty acids (NEFA) plays an important role in the pathogenesis of cardiovascular disease (CVD). Individual circulating NEFA have achieved less scientific focus in relation to atherosclerosis compared to total NEFA. A strong link exists between chronic inflammation and nutritional markers in hemodialysis (HD) patients. Biological effects of n-3 fatty acids have been shown in studies of animals and humans in which anti-inflammatory and antithrombotic effects could be of relevance in HD patients. It is widely accepted that dietary intake of n-3 PUFA or fish and fish oil as their dietary sources have beneficial health effects among sick and healthy individuals. A great number of clinical studies have shown that intake of n-3 fatty acids reduces incidence of cardiovascular events and play an important role in the prevention of inflammatory and autoimmune disorders.[11,12,13,14]

HD patients have impaired essential fatty acid status (EFA), as well as high level of oxidative stress, and very often are malnourished with a high risk for developing cardiovascular pathology.

First aim of our study was to establish the association between markers of inflammation and atherosclerosis with parameters of malnutrition in HD patients. Second aim of this study was to determine the baseline status of EFA in middle-aged hemodialysis patients and within the average middle-aged population in Serbia (healthy control), using the FA composition of erythrocyte total lipids as a reliable parameter of cellular EFA status.Baseline values confirmed the presence of EFA deficiency in HD patients in comparison with the healthy control.

Methods

The University Clinical Research Ethics Committee approved the study. All 42 haemodialysis patients, recruited from our unit gave written informed consent. They had been treated with hemodialysis three times a week for at least 6 months. The patients with diabetes, acute infections and malignancy were excluded from the study. The duration of dialysis was defined as the number of months from initiation of chronic hemodialysis to the time of laboratory data collection. All blood samples were obtained immediately before the dialysis session after an 8-hour fast on a mid-week dialysis session. Plasma levels of albumin, transferin, interleukin-6 (IL-6), tumor necrosis factor – alpha (TNF-α), high sensitivity C-reactive protein (hs-CRP), total cholesterol, triglyceride, HDL-cholesterol, and LDL cholesterol were measured. Plasma albumin, transferrin and lipids were measured by routine laboratory methods. Plasma hs-CRP level, TNF-α and IL-6 were measured by enzyme-linked immunosorbent assay kits (ELISA). Body mass index (BMI) was recorded. Lean body mass (LBM), was measured by bioelectric impedance (BIA). Mid-arm muscle circumference (MAMC) was recorded too.We used standard Doppler echo examinations to determine plaque on carotid arteries. End diastolic pressure (EDD) and ejection fraction (EF) were measured by echocardiography.

The total fatty acids composition was determined by gas chromatography (GS) in the methyl esters form.GS analysis was performed on a VARIAN chromatographer, Model 1400.

All statistical analysis was made with Stat 6.0 software package on MS Windows (Statsoft Inc., Novi Sad Serbia). The t-test and Mann-Whitney test were applied for statistical comparisons between groups and for statistical comparisons before and after treatment in HD patients. Pearson s correlation analysis was used to study correlations between variables.

Statistical multivariate factor analysis of the results was carried out to investigate all parametes. The objective of factor analysis was to reveal a simple underlying structure that is presumed to exist within a set of multivariate observations.

Results

The main results of the patients of each tertile are shown on the Table 1. HD patients of the 3rd tertile, representing 30% of the whole group, presented LBM levels indicative of a well nourished group. These patients presented BMI and other nutritional parameters, significantly higher than that of the patients in the 1st tertile. IL-6 and TNF-α were higher in the patients in the 1st tertile and these differences were significantly according to 3st tertile. Patients with a lower LBM values had decreased BMI, MAMC, albumin level and transferrin level than those in the higher tertile. Malnourished patients had significantly increased cardiovascular disease and carotid plaques. Risk of metabolic and cardiovascular disorder was greater when serum albumin concentration was lower.

Significant positive correlation between TNF-alpha and LDL/HDL cholesterol was found in the whole group of HD patients (Table 2). Parameter of inflammatory process, hs-CRP was positive correlation with atherosclerotic and cardiovascular parameters. There were negative correlation with hs-CRP and albumin and HDL-cholesterol level (Table 3).

There was negative correlation between mid-arm muscle circumference (MAMC) and albumin level and saturated fatty acids. There was positive correlation between diastolic blood pressure and n-3PUFA (Table 6 and 7).

The results showed the total content of the n-3 fatty acids (3.02±0.84 to 3.91±1.01), as well as the total content of the polyunsaturated fatty acids (24.2±3.16 to 29.5±3.87), significantly increased after fish oil supplementation in HD patients. The significant decrease was observed in total saturated fatty acids content (45.7±3.06 to 41.6±4.47), while content of total monounsaturated fatty acids was not changed. A decrease in plasma triglyceride but not significantly and a significant increase in plasma HDL-cholesterol concentrations were observed after fish oil supplementation (Table 4).

After 8 weeks of fish oil supplementation, the status of most individual erythrocyte FAs was not significantly changed in healthy control. The only significant changed were the decrease in oleic acid, 20:3 n-9 FAs and the ratio of n-6/n-3. Noticed increase in EPA and DHA was not statistically significant. After fish oil supplementation concentrations of plasma trigliceride and total cholesterol were significant decreased, but no significant effects in plasma HDL-cholesterol concentration was observed (Table 5).

Results of the factor analysis indicated 4 latent factors that explained 66.94% of the variance within all investigated parameters (Table 8). The 1st factor of all parameters explained 21.96% of the variance. The highest factor loadings were found at the positive end for triglyceride, cholesterol and LDL cholesterol level as the atherogenic factor. Significant factor loadings for the 2nd factor, with 16.51% of the explained variance, had the components of IL-6 and TNF-alpha at the positive end for these parameters, as inflammatory parmeters. The 3rd factor, with 14.76% of the explained variance, had one component with high factor loadings, namely, LBM as a nutritional parameter parameter. The presence of factor 4 with 13.71% of explained varience had both hs-CRP level and albumin level as factors at the positive ends and these are two parameters indicative of an inflammatory process and malnutrition.

Table 1.Comparison between HD groups according LBM tertiles

	1st tertile	3st tertile	p
Number	19	23	
BMIkg/m^2	22±3.4	24±1.97	0.04
MAMC cm	23.9±4.7	25.7±3.1	0.04
Albumin g/L	31±3.01	33±3.21	0.05
Transferin pg/ml	1.42±0.29	1.75±0.32	NS
TNF-α pg/ml	2.08±1.65	1.95±1.53	NS
IL-6 pg/ml	4.50±4.19	3.41±3.39	0.03
Carotid plaques %	62	39	≤0.05
CVD %	89	72	≤0.05

Table 2. Correlation coefficient between TNF-α and LDL/HDL-cholesterol according LBM tertiles

	1st tertile		3rd tertile	
TNF-alpha	r =0.33	p<0.05	r=0.32	p<0.05

Table 3. Correlation between hsCRP and some parameters

Correlation coefficient (r) p-value
Carotid plaque % 0.34 < 0.05 End diastolic diameter (EDD) 0.34 < 0.05 Albumin - 0.24 0.02 HDL-cholesterol - 0.19 0.09

Table 4. Changes in content fatty acids before and after supplementation (%) in HD patients

Fatty acids	Before supplementation	After supplementation	p
Σ SFA	45.7 ± 3.06	41.6 ± 4.47	< 0.05
Σ MUFA	25.6 ± 2.67	25.7 ± 3.40	NS*
Σ PUFA	24.2 ± 3.16	29.5 ± 3.87	< 0.05
Σ n-3	3.02 ± 0.84	3.9 ± 1.01	< 0.05
Σ n-6	21.1 ± 3.12	25.6 ± 3.71	< 0.05
n-6/n-3	7.68 ± 2.94	6.97 ± 2.04	NS*

NS*=no significant.

Table 5. Content fatty acids in phospholipid of membrane erythrocyte before and after supplementation in control group

Fatty acids	Before supplementation	After supplementation	p
Σ SFA	41.3 ± 2.73	41.9 ± 3.41	NS*
Σ MUFA	24.2 ± 2.56	21.4 ± 1.84	< 0.05
Σ PUFA	31.7 ± 4.55	33.3 ± 2.61	NS*
Σ n-3	4.5 ± 1.85	6.0 ± 1.53	< 0.05
Σ n-6	27.2 ± 3.60	27.3 ± 1.95	NS*
n-6/n-3	7.8 ± 5.02	4.57 ± 1.04	< 0.05

NS*=no significant.

Table 6. Correlation between serum albumin level and PUFA

DHA er 22:6(n-3)	r = 0.415	p = 0.05
AA er 20:4(n-6)	r = 0.643	p = 0.002

Table 7. Correlation between diastolic blood pressure and PUFA

DHA er 22:6(n-3)	r = -0.409	p = 0.05
AA er 20:4(n-6)	r = -0.4110	p = 0.06

Table 8. Results of multivariate factor analyses in investigated parameters

STAT FACTOR ANALYSIS	Factor Loadings (Varimax raw) Extraction: Principal components (Marked loadings are >0.7)			
Variable	Factor 1	Factor 2	Factor 3	Factor 4
triglyceride	0.846*	-0.036	0.134	0.014
cholesterol	0.754*	-0.079	0.209	0.154
HDL-cholesterol	-0.401	-0.229	-0.444	0.203
LDL-cholesterol	0.790*	0.210	-0.061	0.047
hs-CRP	0.105	0.044	0.291	0.805*
albumin	-0.034	-0.082	0.308	-0.755*
TNF-alpha	0.162	0.858*	0.004	0.196
IL-6	-0.095	0.836*	0.026	-0.097
transferin	0.202	-0.278	0.675	-0.121
LBM	-0.186	0.153	0.797*	0.144
Expl.Var Prp.Totl	2.195 0.219	1.650 0.165	1.475 0.147	1.371 0.137

The asterisk (*) is meant $P \leq 0.05$ when compared with the mean baseline values.

Table 9 shows that systolic blood pressure (SBP) has a significantly positive correlation with BMI and arachidonic acid and SBP.

Table 9. Correlation matrix between variables

	BMI	Kt/V	SBP	DBP	20:2(n:6)	20:3(n-6)	20:4(n-6)	20:5(n-3)	22:5(n-6)	22:5(n-3)	22:6(n-3)
BMI	1										
Kt/V	0.05245	1									
SBP	0.3429859*	-0.28288	1								
DBP	0.120363	-0.08091	0.79498187*	1							
20:2(n:6)	0.258895	0.023627	0.31587441	0.283477	1						
20:3(n-6)	-0.17542	-0.0093	-0.1638159	0.033166	0.169577	1					
20:4(n-6)	0.220512	-0.12769	0.42942105*	0.280473	0.248284	0.144521	1				
20:5(n-3)	-0.08249	-0.1973	0.01869124	-0.05628	-0.06748	0.358487*	0.068631	1			
22:5(n-6)	0.218991	-0.02948	0.00629863	-0.07456	-0.05616	-0.12298	0.214241	-0.26313	1		
22:5(n-3)	-0.03521	-0.19221	-0.1152159	-0.16658	-0.04875	-0.07349	0.016512	0.072813	-0.11062	1	
22:6(n-3)	0.144039	0.230754	0.23122294	0.185359	0.223635	0.298395	0.130309	0.17112	0.027636	0.22809	1

The asterisk (*) is meant $P \leq 0.05$ when compared with the mean baseline values.

SBP=systolic blood pressure; DBP=diastolic blood pressure; Kt/V=adequacy dialysis; BMI=body mass index

Discussion

This cross sectional study confirms that atherosclerotic risk is presents in high percentage of HD patients. The characteristics of these patients, such as elevated plasma lipids, poor nutritional parameters (dry weight, lean body mass and BMI), high percent inflammation and vascular disease (presented as macroangiopathy), were found to be significant predictors of mortality and comorbidity. Many studies have examined inflammatory markers in the process of endothelial dysfunction. Initiation of the inflammatory process in ESRD can be associated with an elevation of serum acute-phase protein, including CRP and some proinflammatory cytokines such as IL-6 and TNF-α. IL-6 is reported to play a central role in the pathophysiology of the adverse effects of inflammation in ESRD patients. Increased activation of inflammatory cytokines, such as IL-6 and TNF-α,may cause muscle breakdown and hypoalbuminemia and may be involved in atherogenesis. [5,6,7,10]

It is also evident that most of the classical risk factors associated with a tendency to atherosclerotic disease in the general population also play a role in HD patients. Two of these factors, malnutrition and inflammation, also were identified in many clinical studies. It is well recognized that malnutrition and hypoalbuminemia are important predictive factors of mortality in patients with ESRD. In addition, high concentrations of acute-phase proteins (CRP) are strongly associated with death within 1 year in pre-dialysis patients and in dialysis as well. In our study also there was association between CRP and albumin level. Fluid overload and congestive heart failure may be other contributors to increased inflammatory responses in HD patients.[11,12]

It is now known that chronic inflammation increases the risk of premature atherosclerosis and cardiovascular disease in patients with chronic kidney disease. Correction of nutritional status in HD patients improve inflammatory state and decrease in plasma concentration of inflammatory markers and other parameters which contribute of cardiovascular disease.

Malnourished dialysis patients are more often with cardiovascular disease and carotid plaques and have elevated inflammatory cytokines which all may increase the risk of atherosclerotic vascular disease. These data suggested that inflammatory markers in malnourished HD patients will identify patients at high risk of comorbidity and mortality. The degree of inflammation correlates with disease severity.

Inflammation was associated with indices of malnutrition and atherogenic lipids and may contribute to the high frequency of cardiovascular disease in HD patients. End stage renal disease leads to a reduction in lean body mass, which contributes poor the outcome in HD population.

The accelerated atherosclerotic process of ESRD may involve several interrelated conditions, such as oxidative stress, endothelial dysfunction, vascular calcification, and inflammation. Studies in the general population and in ESRD patients have demonstrated a significant association between inflammatory markers and CVD.

In this study we have shown that HD patients had reduce PUFAs in the red cell phospholipid. This study has demonstrated high prevalence of malnutrition, inflammation, carotid plaques and cardiovascular disease. Malnourished dialysis patients are more often with cardiovascular disease and carotid plaques and have elevated inflammatory cytokines which all may increase the risk of atherosclerotic vascular disease.

In the present study, four significant findings were observed: (i) reduced content of essential fatty acids in red cell membrane phospholipids; (ii) reduced levels of some nutritional parameters (indicative of malnutrition) and increased plasma inflammatory markers (IL-6 and TNF-a); (iii) IL-6 and TNF-a correlated with some essential fatty acids; and (iv) 2 months of omega-3 treatment normalizedmost of these biochemical changes. These data indicate that PUFA deficiency may be associated with malnutrition and inflammation in those patients and their levels after treatment may reflect clinical outcome.

Phospholipids play a main role in the membrane function. The length and degree of unsaturated fatty acids are determinants of enzymes activity. The activation of peroxisome proliferator activated receptors (PPARs) might be a possible mechanism for down regulaton of inflammatory markers with supplementation of n-3 fatty acids have been shown in many studies of animals and humans in which anti-inflammatory effects theoretically could be of relevance in patients with renal disease. These data indicate that PUFAs deficiency may be associated with increase inflammatory cytokines and their levels after treatment may reflect on better clinical outcome.

In summary, this cross-sectional study in HD patients has demonstrated high prevalence of malnutrition, inflammation, carotid plaques and cardiovascular disease. The results suggest that inflammation is an important process in promoting atherosclerosis in malnourished haemodialysis patients.

The red cell membrane is an example of plasma membrane. It is very similar to the vascular endothelium. A loss of PUFA content in membrane phospholipid would reduce liquidity and the mechanical properties normal cell membranes. Such a defect in the endothelial membrane could cause vascular dysfunctions as demonstrated in rats. This suggests that the pathology of the inflammatory process is associated with the incorporation of EPA and DHA into membrane phospholipids [18,19]

The importance of lipid level as a risk factor in HD patients is more complex and depends on other factors, such as nutritional parameters and inflammatory processes. IL-6 is a central mediator of cardiovascular risks associated with many diverse conditions.[20,21,22]

The pilot study confirmed the presence of an essential fatty acid deficiency in HD pts. There was positive correlation between diastolic blood pressure and PUFA-18:3(n-3). There were positive correlation between MAMC and PUFA-22:6(n-3) and 18:3(n-3) too. There was negative correlation between MAMC and albumin level and SFA.[23,24,25]

Peck suggests that HD patients may have increased prostaglandin E2 (PGE2) values and fatty acid profiles indicative of EFAD. Sinziger and Leitner have shown that the metabolism of arachidonic acid shifted from the cyclooxygenase pathway to the lipooxygenase pathway in renal patients. Higher level of eicosapentaenoic acid (EPA) and docosahexaenoic acid (DHA) block the lipooxygenase pathway of eicosanoid production, pushing the arachidonic acid metabolism to cyclooxigenase pathway. This pathway leads into increased level of prostaglandin E2 (PGE2) which acts less inflammatory than leukotriene B4 [26,27].

Conclusion

Malnutrition is of greater prognostic importance in the survival of HD pts. Malnutrition is associated with number of metabolic and vascular abnormalities. These factors include hypoalbuminemia, dyslipidemia with raised triglyceride concentrations and low-density lipoprotein, insulin resistance and high concentrations of acute-phase proteins. Low serum albumin concentration, usually used as an index of malnutrition, are highly associated with increased mortality risk in dialysis patients. However, serum albumin is influenced by factors other than malnutrition, and high concentrations of acute-phase proteins, such as C-reactive protein (CRP), are correlated with low serum albumin in malnourished patients of dialysis.

The results indicate that new treatment is needed to improve the nutritional status, quality of life and decreased co morbidity and mortality in HD pts. Cellular basis of pathogenetic factors in malnutrition is, however, unclear. It is now recognized that oxidative stress and inflammatory cytokine are aggravated nutritional status in those patients.

References

[1] Stenvinkel P. Inflammatory and atherosclerosis interactions in the depleted uremic patients. *Blood Purif.* 2001; 19: 53–61.

[2] Zimmermann J, Herrlinger S, Pruy A *et al.* Inflammation enhances cardiovascular risk and mortality in hemodialysis patients. *Kidney Int.* 1999; 55: 648–58.

[3] Bologa RM, Levine DM, Parker TS *et al.* Interleukin-6 predicts hypoalbuminemia, hypocholesterolemia, and mortality in hemodialysis patients. *Am. J. Kidney Dis.* 1998; 32: 107–14.

[4] Kaizu Y, Kimura M, Yoneyama T *et al.* Interleukin-6 may mediate malnutrition in chronic hemodialysis patients. *Am. J. Kidney Dis.*1998; 31: 93–100.

[5] Stenvinkel P, Heimburger O, Jogestrand T. Elevated interleukin-6 predicts progressive carotid atherosclerosis in dialysis patients: Association to chlamydia pneumoniae seropositivity. *Am. J.Kidney Dis.* 2001; 00: 000–000.

[6] Ross R. Atherosclerosis: An inflammatory disease. *N. Engl. J. Med.*1999; 340: 115–26.

[7] Ebisuri C, Tsujinaka T, Morimoto T *et al.* Interleukin-6 induces proteolysis by activating intracellular proteases (cathepsin B and L, proteasome) in C2C12 myotubes. *Clin. Sci.* 1995; 89: 431–9.

[8] Tsujinaka T, Fujita J, Ebisuri C *et al.* Interleukin-6 receptor antibodyinhibits muscle atrophy and modulates proteolytic systems in interleukin-6 transgenic mice. *J. Clin. Invest.* 1996; 97: 244–9.

[9] King AJ, Kehayas JJ, Roubenhoff R *et al.* Cytokine production and nutritional status in hemodialysis patients. *Int. J. Artif. Organs* 1998; 21: 4–11.

[10] Stenvinkel P, Heimburger O, Paultre F *et al.* Strong association between malnutrition, inflammation, and atherosclerosis in chronic renal failure. *Kidney Int.* 1999; 55: 1899–911.

[11] Li YP, Lecker SH. Chen Y, Waddell ID, Goldberg AL, and Reid MB.TNF-alpha increases ubiquity-conjugating activity in skeletal muscle by up-regulating UbcH2/E220k. *FASEB J* 2003; 17:1048-1057.

[12] Lowrie EG and Lew NL. Death risk in hemodialysis patients: The predictive value of commonly measured variables and an evaluation of date rate differences between facilities. Am J Kidney Dis 1990;15:458-482.

[13] Simopoulos AP. Omega-3 fatty acids in health and disease and in growth and development. *Am. J. Clin. Nutr.* 1991; 54: 438–63.

[14] Burr ML, Fehily AM, Gilbert JF *et al.* Effect of change in fat, fish and fibre intakes on death and myocardial infarction: Diet and reinfarction trial (DART). *Lancet* 1989; 2: 757–61.

[15] de Lorgeril M, Renaud S, Mamelle N *et al.* Mediterraneana-linolenic acid-rich diet in secondary prevention of coronary heart disease. *Lancet* 1994; 343: 1454–9.

[16] Siscovick DS, Raghunathan TE, King I *et al.* Dietary intake and cell membrane levels of long-chain n-3 polyunsaturated fatty acids and the risk of primary cardiac arrest. *JAMA* 1995; 274: 1363–7.

[17] Sellmayer A, Witzgall H, Lorenz RL, Weber PC. Effect of dietary fish oil on ventricular premature complexes. *Am. J. Cardiol.* 1995;76: 974–7.

[18] Koorts AM, Viljoen M, Kruger MC. Red blood cell fatty acid profile of chronic renal failure patients receiving maintenance haemodialysis treatment. *Prostaglandins Leukot. Essent. Fatty Acids*2002; 67: 13–18.

[19] G.Perunicic-Pekovic, S.Pljesa, Z.Rasic-Milutinovic, S.Stankovic, M.Ilic, and R.Maletic. Inflammatory cytokines and malnutrition as related to risk for cardiovascular disease in hemodialysis patients Can J Physiol. Pharmacol 86:205-209 (2008)

[20] G. Peruničić-Peković, Z. Rašić, S. Plješa, S. Šobajić, I Djuričić, R Maletić, D Ristić-Medić. Effect of n-3 fatty acids on nutritional status and inflammatory markers in haemodialysis patients. Nephrology 2007; 12, 331-336

[21] V.Ristić, V. Tepšić, D. Ristić-Medić, G Peruničić,.Z. Rašić, M. Poštić, A. Arsić, V. Blaženčić, and G. Ristić. Plasma and erythrocyte Phospholipid fatty acids composition in Serbian hemodialysis patients. Renal Failure 2006; 28:211-216

[22] Z. Rašić-Milutinović, G. Peruničić-Peković, S. Plješa, Z. Gluvić, S. Sobajic, I Djuric, D. Ristić. Effects of N-3 PUFAs supplementation on insulin resistance and inflammatory biomarkers in hemodialysis patients. *Renal failure* 2007; 29: 321-329

[23] I.Djuričić, S. Šobajić, G.Peruničić, M.Stojanov, Z. Rašić. Consumption of fish oil supplement alters erytrocyte fatty acid composition in overweight, hypercholesterolemic, middle-aged Serbians. *Nutrition Research* 2007; 27: 529-534

[24] G.Peruničić-Peković, Z.Rašić, V. Ristić, S. Plješa, LJ. Lambić, V. Djurković. Interactions between nutritional parameters and polyunsaturated fatty acid (PUFA) profiles in plasma phospholipids in hemodialysis (HD) patients. *Cardionephrology* 8, 10th Assisi European Meeting on Cardionephrology, Assisi, April 1-3,2004, 333-335

[25] Peck LW. Essential fatty acid deficiency in renal failure: Can supplements really help? *J.Am Diet.Assoc.* 1997; 97 (Suppl 2): S150-53.

[26] Sinzinger M, Leitner C. Eicosanoids of platelet and vascular wall in chronic renal insufficiency. *Am. J. Nephrol.* 1987; 7: 211-30.

In: Fish Consumption and Health
Editors: George P. Gagne and Richard H. Medrano
ISBN 978-1-60741-151-2

Chapter VI

Omega-3 Dietary Supplements and the Risk of Cardiovascular Events: A Systematic Review

***Paul E. Marik**[*,1] **and Susan Baik**[2]*
[1]Division of Pulmonary and Critical Care Medicine
Thomas Jefferson University, Philadelphia, PA
[2] Baik Nutrition Center, Voorhees, NJ

Abstract

Background: Epidemiologic data suggest that omega-3 fatty acids derived from fish oil reduce cardiovascular disease.

Objective: To assess whether dietary supplements of eicosapentaenoic acid (EPA) and docosahexaenoic acid (DHA) alter cardiovascular deaths and cardiovascular events

Data Sources: MEDLINE, Embase, the Cochrane Database of Systematic Reviews and citation review of relevant primary and review articles.

Study Selection: Prospective, randomized, placebo-controlled clinical trials that evaluated clinical cardiovascular endpoints (cardiovascular deaths and all cardiovascular events) in patients randomized to EPA/DHA or placebo. We only included studies that used dietary supplements of DHA/EPA which were administered for at least one year.

Data Extraction: Data were abstracted on study design, study size, type and dose of omega-3 supplement, cardiovascular events and duration of followup. Studies were grouped according to the risk of cardiovascular events (high risk and moderate/low risk). Meta-analytic techniques were used to analyze the data.

Data Synthesis: We identified nine studies that included 31 996 patients. Dietary supplementation with omega-3 fatty acids significantly reduced the risk of cardiovascular deaths (OR 0.8; 0.71-0.95, p=0.007). The mortality benefit was largely due to the studies

* Address for Correspondence: Paul Marik, MD, FCCP, FCCM, Chief, Pulmonary and Critical Care Medicine,Thomas Jefferson University, 834 Walnut Street, Suite 650, Philadelphia, PA, 19107, 215 955-0830 (Fax), paul.marik@jefferson.edu.

which enrolled high risk patients. Similarly, dietary supplementation with omega-3 fatty acids significantly reduced the risk of all cardiovascular events (OR 0.82; 95% CI 0.74 to 0.90, p< 0.0001). This benefit was present in both high and low risk groups. Meta-regression failed to demonstrate a relationship between the daily dose of omega-3 fatty acid and clinical outcome.

Conclusions: Dietary supplementation with omega-3 fatty acids should be considered in patients at risk for cardiovascular disease.

Keywords: omega-3, dietary supplements, angina, myocardial infarction, death, cardiovascular disease

Epidemiologic data suggest that omega-3 fatty acids derived from fish oil reduce cardiovascular disease.[1-4] Greenland Eskimos who consume a diet high in omega-3 fatty acids have been noted to have a very low incidence of coronory artery disease (CAD). [5,6] Moreover, several prospective epidemiologic studies have reported a significant association between fish intake and a lower risk of CAD. However, the results of prospective interventional studies and meta-analyses have produced mixed results.[7-10] A number of factors likely contribute to this finding. Most importantly, the benefits of the change in lipoprotein composition as well as the anti-inflammatory and anti-atherogenic properties of a diet high in omega-3 fatty acids may take considerable time to manifest. Studies and meta-analyses which have included patients followed for less than a year may be unable to detect a clinical benefit.[10] Furthermore, while compliance with a dietary supplement (capsules) may be greater than dietary interventions, it is not clear that both intervention have similar effects. Many systematic reviews have included both dietary interventions and dietary supplements; this is likely to lead to significant heterogeneity.[7] Equally important is the fact that different formulations with different dosages have been used. Alpha-linolenic acid (ALA; 18:3) is a plant omega-3 fatty acid which is poorly converted to eicosapentaenoic acid (EPA) and docosahexaenoic acid (DHA) and therefore unlikely to have the cardioprotective effects of fish oil. In addition, a number of studies have been recently published which were not included in previous meta-analyses..[7-10] The goal of this study was to perform a meta-analysis of prospective randomized controlled trials that evaluated the cardioprotective effects of dietary supplements of DHA/EPA when administered for at least one year.

Methods

Identification of Trials

Our aim was to identify all relevant clinical trials that compared the incidence of cardiovascular deaths and combined cardiovascular events (deaths, unstable angina, myocardial infarction, sudden death) in patients who were randomized to receive dietary supplements of EPA/DHA with that of patients who received placebo. We only included studies that used dietary supplements of DHA/EPA which were administered for at least one year. There was no restriction as to the type of patient, clinical setting or language in which the study was reported. We used a multi-method approach to identify relevant studies for this review. Both authors independently searched the National Library of Medicine's MEDLINE database for relevant studies published from 1966 to December 2007 using the following Medical Subject Headings (MeSH) and keywords; omega 3 fatty acids, eicosapentaenoic acid, docosahexaenoic acid and cardiovascular disease, angina, myocardial infarction, sudden cardiac death and randomized controlled trial (publication type). In addition we searched Embase and the Cochrane Database of Systematic Reviews. Bibliographies of all selected articles and review articles that included information on omega-3 fatty acids and cardiovascular disease. This search strategy was done iteratively, until no new potential citations were found. We performed this meta-analysis according to the guidelines proposed by the QUOROM group.[11]

Data Extraction and Analysis

Both authors independently abstracted data from all eligible studies using a standardized form. Data were abstracted on study design, study size, study setting, type and dosage of omega-3 fatty acid used and duration of followup. We recorded the method of randomization, blinding and concealment. *A priori* the studies were grouped according to the risk of cardiovascular events as follows i) high risk (recent acute myocardial infarction) and ii) moderate to low risk (primary prevention or secondary prevention in patients with known atherosclerotic disease).

Statistical analysis was performed using Review Manager 5.0 (Cochrane Collaboration, Oxford) and Comprehensive Meta-analysis 2.0 (Biostat, Englewood NJ). We assessed heterogeneity between studies using the Cochran Q statistic with $p \leq 0.10$ indicating significant heterogeneity and the I^2 with suggested thresholds for low (25-49%) moderate (50-74%) and high (>75%) values.[12-15] The fixed effects model was used when insignificant heterogeneity was present and the random effects model when significant heterogeneity was present for the end-point of interest. For all analyses we considered $p \leq 0.05$ (two sided) as significant. Summary effects estimates are presented as odds ratio (OR) with 95% confidence intervals. Meta-regression between the treatment effect and the dose of fish oil was performed using Comprehensive Meta-analysis software.

Results

The initial search strategy generated 203 citations; of these 65 were review articles, commentaries or observational studies and were excluded from further analysis. A further 129 studies were excluded as they did not report the outcomes of interest or the duration of the study was less than 1 year. Nine studies met our inclusion criteria and were included in the meta-analysis.[16-20,20,21,21-24] All studies used random allocation and all except two were blinded with concealment of allocation.[19,24]These studies are summarized in Table 1.

Table 1. Summary of Studies included in meta-analysis

Author/Year	Country	Type	No Pts	Omega 3	Dose (g)	Control	Duration (yrs)
Nye,1990[16]	N. Zealand	PTCA	73	EPA	2.16	olive oil	1
Sacks,1995[17]	USA	CAD	59	EPA+DHA	4.8	olive oil	2.4
Singh,1997[18]	India	AMI	240	EPA+DHA	1.8	non oil placebo	1
GISSI, 1999[19]	Italy	AMI	11 334	EPA+DHA	0.85	Control +/- Vit E	3.5
Von Schacky, 1999[20]	Germany	CAD	223	EPA+DHA	1.7	mixed fatty acids	2
Leng,1998[21]	Scotland	PVD	120	EPA	0.27	sunflower oil	2
Nilson,2001[22]	Norway	AMI	300	EPA+DHA	3.4	Corn oil	1.5
Leaf,2005[23]	USA	ICD	402	EPA+DHA	2.6	Olive oil	1
Yokoyama,2007[24]	Japan	Hyperchol.	18 645	EPA	1.8	Control	4.6

CAD=coronary artery disease; CABG=coronary artery bypass grafting; AMI=acute myocardial infarction; PVD=peripheral vascular disease.

PTCA=percutaneous transluminal coronary angioplasty; ICD= implanted cardioverter-defibrillator; Hyperchol=hypercholesterolemia; EPA=eicosapentaenoic acid; DHA=docosahexaenoic acid.

Study or Subgroup	Treatment Events	Treatment Total	Control Events	Control Total	Weight	Odds Ratio M-H, Fixed, 95% CI	Year
1.2.1 High Risk							
Singh 1997	14	122	26	118	5.8%	0.46 [0.23, 0.93]	1997
GISSI 1999	291	5666	348	5668	81.4%	0.83 [0.71, 0.97]	1999
Nilson 2001	8	150	8	150	1.9%	1.00 [0.37, 2.74]	2001
Subtotal (95% CI)		**5938**		**5936**	**89.0%**	**0.81 [0.69, 0.94]**	
Total events	313		382				
Heterogeneity: Chi² = 2.72, df = 2 (P = 0.26); I² = 27%							
Test for overall effect: Z = 2.72 (P = 0.007)							
1.2.2 Low Risk							
Nye 1990	0	0	0	0		Not estimable	1990
Sacks 1995	0	31	1	28	0.4%	0.29 [0.01, 7.44]	1995
Leng 1998	2	60	2	60	0.5%	1.00 [0.14, 7.34]	1998
von Schacky 1999	0	112	1	111	0.4%	0.33 [0.01, 8.12]	1999
Leaf 2005	9	200	9	202	2.1%	1.01 [0.39, 2.60]	2005
Yokoyama 2007	29	9326	31	9319	7.6%	0.93 [0.56, 1.55]	2007
Subtotal (95% CI)		**9729**		**9720**	**11.0%**	**0.91 [0.59, 1.39]**	
Total events	40		44				
Heterogeneity: Chi² = 0.93, df = 4 (P = 0.92); I² = 0%							
Test for overall effect: Z = 0.44 (P = 0.66)							
Total (95% CI)		**15667**		**15656**	**100.0%**	**0.82 [0.71, 0.95]**	
Total events	353		426				
Heterogeneity: Chi² = 3.95, df = 7 (P = 0.79); I² = 0%							
Test for overall effect: Z = 2.71 (P = 0.007)							

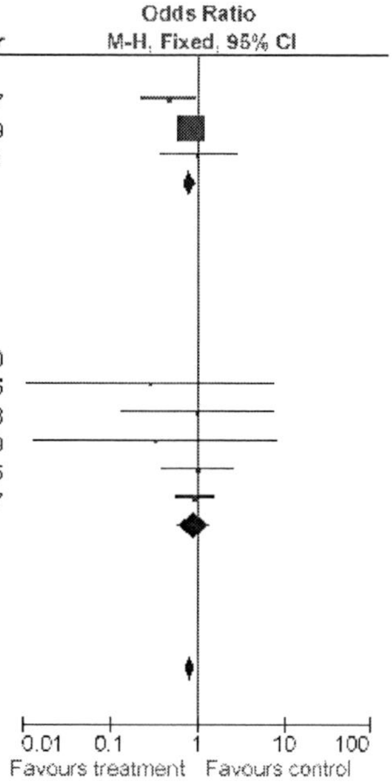

Figure 1. Effect of omega-3 fatty acid supplementation on cardiovascular deaths, stratified by risk classification (odds ratio and 95% confidence interval).

In all 31,396 patients were included in the nine studies. The study size varied from 59 to 18,645 patients. Three studies enrolled high risk patients (AMI)[18,19,22], five studies were secondary prevention studies in patients with CAD,[16,17,20,21,23] and one was a primary prevention study in patients with hypercholesterolemia.[24] The combined dosage of DHA/EPA varied from 0.27 to 4.8g/day (mean 2.0 g/day) and the duration of followup varied from 1 to 4.6 years (mean 2.1 years). Three studies used olive oil as a control.[16,17,23]

Dietary supplementation with omega-3 fatty acids significantly reduced the risk of cardiovascular deaths (OR 0.8; 0.71-0.95, $p=0.007$) (see Figure 1). There was insignificant heterogeneity between studies ($p=0.79$, $I^2=0\%$). The mortality benefit was largely due to the studies which enrolled high risk patients. Similarly, dietary supplementation with omega-3 fatty acids significantly reduced the risk of all cardiovascular events (OR 0.81; 95% CI 0.74 to 0.90, $p< 0.0001$)(see Figure 2). There was insignificant heterogeneity between studies ($p=0.3$, $I^2=16\%$). This benefit was present in both high and low risk groups. Meta-regression revealed no association between the treatment effect and the dose of fish oil (see Figure 3).

Discussion

This meta-analysis demonstrates that dietary supplementation with omega-3 fatty acids for a year or longer significantly reduces the risk of cardiovascular deaths and all cardiovascular events. We were however unable to demonstrate a dose response. This results of our meta-analysis are similar to those of Butcher et al, however, these authors included randomized trials of dietary intervention as well as omega-3 fatty acids supplements and included trials with a followup of at least 6 months. [10] He and colleagues performed a meta-analysis of cohort studies to examine the association between fish intake and CHD mortality.[8] These authors reported a significant reduction in cardiac deaths associated with increasing consumption of fish, with a clear dose response effect. The study by Nilsen and colleagues was the only study included in our meta-analysis which did not demonstrate a clinical benefit from omega-3 fatty acids.[22] This observation may be explained by the fact that the participants in this study resided in a costal area with a diet rich in fish products. Omega-3 supplementation may not have additional benefits in patients who consume fish on a regular basis. However, omega-3 fatty acid supplementation was beneficial in the GISSI-Prevenzione trial in which the majority of subjects reported eating at least one fish serving a week.[19]

Study or Subgroup	Treatment Events	Treatment Total	Control Events	Control Total	Weight	Odds Ratio M-H, Fixed, 95% CI	Year
2.2.1 High Risk							
Singh 1997	30	122	41	118	3.4%	0.61 [0.35, 1.07]	1997
GISSI 1999	424	5666	485	5668	48.6%	0.86 [0.75, 0.99]	1999
Nilson 2001	75	150	71	150	3.8%	1.11 [0.71, 1.75]	2001
Subtotal (95% CI)		**5938**		**5936**	**55.8%**	**0.87 [0.76, 0.98]**	
Total events	529		597				
Heterogeneity: Chi² = 2.65, df = 2 (P = 0.27); I² = 24%							
Test for overall effect: Z = 2.22 (P = 0.03)							
2.2.2 Low Risk							
Nye 1990	9	36	16	37	1.3%	0.44 [0.16, 1.18]	1990
Sacks 1995	7	31	11	28	1.0%	0.45 [0.15, 1.40]	1995
Leng 1998	8	60	11	60	1.0%	0.69 [0.25, 1.85]	1998
von Schacky 1999	2	112	7	111	0.7%	0.27 [0.05, 1.33]	1999
Leaf 2005	57	200	78	202	6.0%	0.63 [0.42, 0.96]	2005
Yokoyama 2007	262	9326	324	9319	34.1%	0.80 [0.68, 0.95]	2007
Subtotal (95% CI)		**9765**		**9757**	**44.2%**	**0.75 [0.65, 0.87]**	
Total events	345		447				
Heterogeneity: Chi² = 4.76, df = 5 (P = 0.44); I² = 0%							
Test for overall effect: Z = 3.82 (P = 0.0001)							
Total (95% CI)		**15703**		**15693**	**100.0%**	**0.81 [0.74, 0.90]**	
Total events	874		1044				
Heterogeneity: Chi² = 9.47, df = 8 (P = 0.30); I² = 16%							
Test for overall effect: Z = 4.18 (P < 0.0001)							

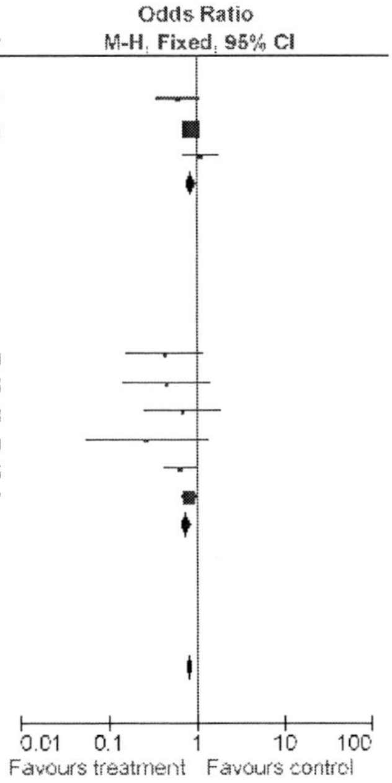

Figure 2. Effect of omega-3 fatty acid supplementation on all cardiovascular events, stratified by risk classification (odds ratio and 95% confidence interval).

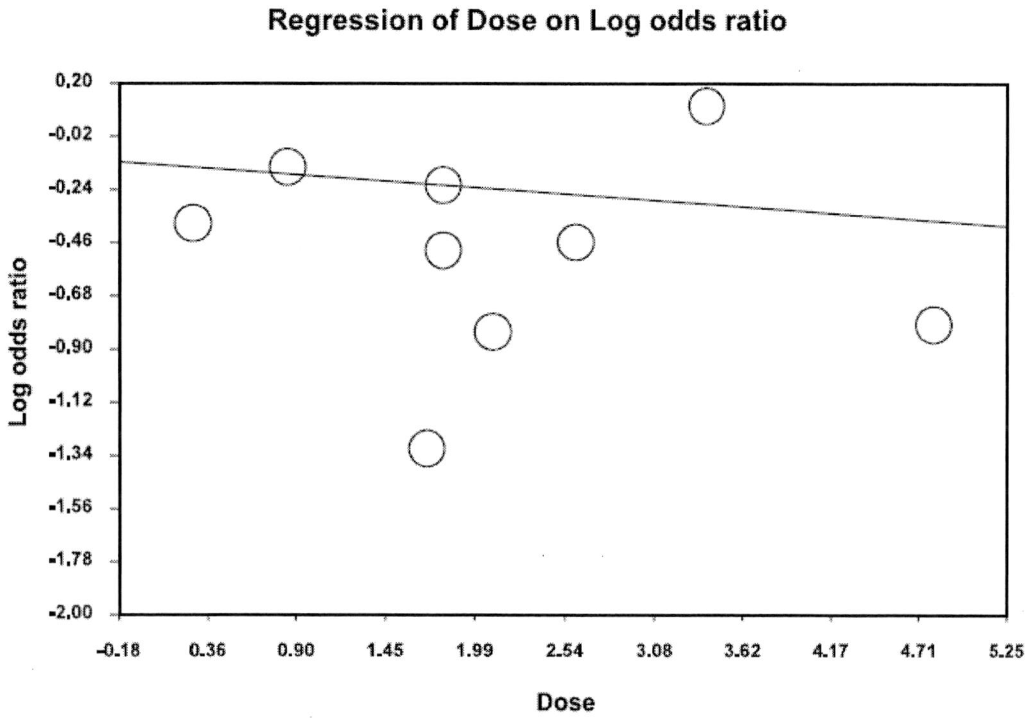

Figure 3.Regression of dose of omega-3 fatty acid on log of odds ratio.

The study by Burr and coauthors was excluded from our meta-analysis.[25] In this study patients were randomized to i) eat two portions of oily fish or to a fish oil supplement as a partial or total substitute, ii) a fruit and vegetable diet, iii) a combination of both of these diets and iv) a "sensible eating" dietary advice control group. The results of the study suggested that men advised to eat fish, particularly those supplied with fish oil capsules had a higher risk of cardiac death. The findings of this study are at odds with the results of our meta-analysis and the data from epidemiological and cohort studies demonstrating a cardio-protective effect of omega-3 fatty acids.[1-4] This study has numerous methodological issues which may have accounted for this finding. Furthermore, it is possible that the fish eaten in this study had a high mercury contnet. Fish oils may contain high levels of various toxic compounds such as mercury, dioxins and polychlorinated biphenyls. These compounds are fat soluble and accumulate in the body over time. Salonen and coauthors demonstrated that the intake of freshwater fish high in mercury was associated with an increased risk of acute myocardial infarction and death from cardiovascular disease.[26] Mercury reduces intracellular and plasma antioxidant capacity and promotes peroxidation of cell membranes and lipoproteins. In addition, mercury compounds promote platelet aggregation and stimulate the proliferation of arterial smooth muscle cells.[27]These effects of mercury may promote atherosclerosis. These data suggest that the beneficial effects of omega-3 fatty acids may be negated by fish with a high mercury content. This effect does not apply to pharmaceutical grade fish oil supplements which are highly purified with the removal of mercury and other contaminants.

The human body evolved on a diet in which there was a balance between omega-6 and omega-3 fatty acids (i.e a ratio of approximately 1:1). Today the ratio of these lipids in the diet is approximately 14:1. The net result is an exaggeration of omega-6 lipid effects. Numerous studies clearly indicate that the relative quantities of omega-6 and omega-3 fatty acids in the diet alter a variety of cell functions and disease states. Omega-3 fatty acids have several beneficial cardiovascular properties, including anti-atherothrombotic, antiarrhythmic, anti-inflammatory, antihypertensive and triglyceride lowering. As demonstrated by this meta-analysis and other reports, omega-3 fatty acids reduce the risk of sudden cardiac death as well as fatal and non-fatal myocardial infarction.[3,23,28,29] The protective effect of omega-3 fatty acids may be mediated via a number of mechanisms. Omega-3 fatty acids decrease the production of proatherogenic and proinflammatory proteins in human endothelial cells, reduce endothelial dysfunction by reducing sympathetic overactivity and enhance nitric-oxide mediated vasodilatation.[30,31]They reduce the endothelial inflammatory response by inhibiting monocyte adhesion and by decreasing thromboxane A2, a prostaglandin that induces platelet aggregation and vasoconstriction.[32,33] Platelet-derived growth factor-like proteins from vascular endothelial cells are also suppressed by omega-3 fatty acids, which may attenuate endothelial cell proliferation. Omega-3 fatty acids affect hemostatic markers of atherosclerosis, such as thrombomodulin and von Willebrand factor, as well as reduce insulin resistance. [34]

Omega-3 fatty acids are incorporated into cellular membranes where they influence fluidity and function. This effect is especially important in hormone binding, cell-receptor action, ion channel function, transporter activity, membrane bound enzyme activity, signal transduction, cell-to-cell interactions and release of membrane lipid precursors. The lipid content of the cell membrane affects activity of adenyl cyclase and Na/K ATPase. The antiarrhythmic effects of omega-3 fatty acids may relate to modulation of sodium channels by direct binding of the lipids to channel protein.[35] Omega-3 fatty acids have been demonstrated to alter the transcription of specific genes.[30,36] These effects are mediated by a variety of mechanisms that involve indirect (i.e by eicosanoids, hormones) and direct nuclear effects on genes. Both omega-6 and omega-3 fatty acids modulate expression of genes involved in lipogenesis, glycolysis, production of glucose transporters, inflammatory mediators, early response genes and genes for cell adhesion molecules. Because omega-6 and omega-3 fatty acids possess opposing actions, both classes of lipids are required in proper amounts for optimal growth, development, response to injury and overall health. Recent evidence suggests that decreased production of some inflammatory mediators is mediated at a transcriptional level. EPA and DHA reduce the expression of genes for IL-6, IL-8, IL-1β, vascular cell adhesion molecule-1 (VCAM), intracellular adhesion molecule-1 (ICAM-1), endothelial adhesion molecule and E-selectin.[31]

It should be noted that olive oil was used as the control in three of the studies included in our meta-analysis.[16,17,23] Furthermore, most patients in the GISSI-prevenzione trial were on a Mediterranean diet and taking olive oil.[19] Olive oil is a major component of the Mediterranean diet and has beneficial effects on lipoprotein metabolism and glucose control and decreases oxidative injury, endothelial dysfunction and inflammation.[37-41] This suggests that omega-3 fatty acids have additional benefits over and above olive oil and a Mediterranean diet. Furthermore, the use of olive oil as the control in the three studies noted above may have reduced the "true" benefit of omega-3 fatty acid supplementation.[16,17,23] Most patients in the primary and secondary prevention trials included in this meta-analysis were receiving concomitant treatment with a HMG CoA reductase inhibitor (statin) and aspirin. It is likely that omega-3 fatty acids act synergistically with these drugs. The combination of omega-3 fatty acids and statins have shown beneficial effects on lipid profiles in patients with hyperlipidemia.[42,43]

In summary, the results of this meta-analysis demonstrate that dietary supplementation with omega-3 fatty acids decrease sudden cardiac deaths and all cardiovascular events. These benefits are most apparent in high risk patients. Omega-3 fatty acid supplementation appears to confer additional benefits in patients eating a Mediterranean diet. The optimal daily dose of omega-3 supplementation is unclear, however low dose (1g/day) supplementation is effective and should be considered the preferred dose at this time.

References

[1] Kromhout D, Bosschieter EB, de Lezenne Coulander C. The inverse relation between fish consumption and 20-year mortality from coronary heart disease. *N. Engl. J. Med.* 1985; 312:1205-1209.

[2] Kromhout D, Feskens EJ, Bowles CH. The protective effect of a small amount of fish on coronary heart disease mortality in an elderly population. *Int. J. Epidemiol.* 1995; 24:340-345.

[3] Daviglus ML, Stamler J, Orencia AJ et al. Fish consumption and the 30-year risk of fatal myocardial infarction. *N. Engl. J. Med.* 1997; 336:1046-1053.

[4] Burr ML, Fehily AM, Gilbert JF et al. Effects of changes in fat, fish, and fibre intakes on death and myocardial reinfarction: diet and reinfarction trial (DART). *Lancet* 1989; 2:757-761.

[5] Bang HO, Dyerberg J. Plasma lipids and lipoproteins in Greenlandic west coast Eskimos. *Acta Med. Scand.* 1972; 192:85-94.

[6] Bang HO, Dyerberg J, Nielsen AB. Plasma lipid and lipoprotein pattern in Greenlandic West-coast Eskimos. *Lancet* 1971; 1:1143-1145.

[7] Hooper L, Thompson RL, Harrison RA et al. Omega 3 fatty acids for prevention and treatment of cardiovascular disease. Cochrane Database of Systematic Reviews 2004;CD003177.

[8] He K, Song Y, Daviglus ML et al. Accumulated evidence on fish consumption and coronary heart disease mortality: a meta-analysis of cohort studies. *Circulation* 2004; 109:2705-2711.

[9] Wang C, Harris WS, Chung M et al. n-3 Fatty acids from fish or fish-oil supplements, but not alpha-linolenic acid, benefit cardiovascular disease outcomes in primary- and secondary-prevention studies: a systematic review. *Am. J.Clin. Nutr.* 2006; 84:5-17.

[10] Bucher HC, Hengstler P, Schindler C et al. N-3 polyunsaturated fatty acids in coronary heart disease: a meta-analysis of randomized controlled trials. *Am. J. Med.* 2002; 112:298-304.

[11] Moher D, Cook DJ, Eastwood S et al. Improving the quality of reports of meta-analyses of randomised controlled trials: the QUOROM statement. Quality of Reporting of Meta-analyses. *Lancet* 1999; 354:1896-1900.

[12] Cochran W. The combination of estimates from different experiments. *Biometrics* 1954; 10:101-129.

[13] Berlin JA, Laird NM, Sacks HS et al. A comparison of statistical methods for combining event rates from clinical trials. Stat Med 1989; 8:141-151.

[14] Higgins JP, Thompson SG. Quantifying heterogeneity in a meta-analysis. *Stat. Med.* 2002; 21:1539-1558.

[15] Higgins JP, Thompson SG, Deeks JJ et al. Measuring inconsistency in meta-analyses. *Br. Med. J.* 2003; 327:557-560.

[16] Nye ER, Ablett MB, Robertson MC et al. Effect of eicosapentaenoic acid on restenosis rate, clinical course and blood lipids in patients after percutaneous transluminal coronary angioplasty. *Aust. NZ J. Med*.1990; 20:549-552.

[17] Sacks FM, Stone PH, Gibson CM et al. Controlled trial of fish oil for regression of human coronary atherosclerosis. HARP Research Group. *J. Am. Coll. Cardiol.* 1995; 25:1492-1498.

[18] Singh RB, Niaz MA, Sharma JP et al. Randomized, double-blind, placebo-controlled trial of fish oil and mustard oil in patients with suspected acute myocardial infarction: the Indian experiment of infarct survival--4. *Cardiovasc. Drugs Ther.* 1997; 11:485-491.

[19] Dietary supplementation with n-3 polyunsaturated fatty acids and vitamin E after myocardial infarction: results of the GISSI-Prevenzione trial. *Lancet* 1999; 354:447-455.

[20] von Schacky C, Angerer P, Kothny W et al. The effect of dietary omega-3 fatty acids on coronary atherosclerosis. A randomized, double-blind, placebo-controlled trial. *Ann. Intern. Med*.1999; 130:554-562.

[21] Leng GC, Lee AJ, Fowkes FG et al. Randomized controlled trial of gamma-linolenic acid and eicosapentaenoic acid in peripheral arterial disease. *Clinical Nutrition* 1998; 17:265-271.

[22] Nilsen DW, Albrektsen G, Landmark K et al. Effects of a high-dose concentrate of n-3 fatty acids or corn oil introduced early after an acute myocardial infarction on serum triacylglycerol and HDL cholesterol. *Am. J. Clin. Nutr.* 2001; 74:50-56.

[23] Leaf A, Albert CM, Josephson M et al. Prevention of fatal arrhythmias in high-risk subjects by fish oil n-3 fatty acid intake. *Circulation* 2005; 112:2762-2768.

[24] Yokoyama M, Origasa H, Matsuzaki M et al. Effects of eicosapentaenoic acid on major coronary events in hypercholesterolaemic patients (JELIS): a randomised open-label, blinded endpoint analysis. *Lancet* 2007; 369:1090-1098.

[25] Burr ML, Shfield-Watt PA, Dunstan FD et al. Lack of benefit of dietary advice to men with angina: results of a controlled trial. *Eur. J. Clin. Nutr.* 2003; 57:193-200.

[26] Salonen JT, Seppanen K, Nyyssonen K et al. Intake of mercury from fish, lipid peroxidation, and the risk of myocardial infarction and coronary, cardiovascular, and any death in eastern Finnish men. *Circulation* 1995; 91:645-655.

[27] Kostka B. Potentiation of ADP-induced platelet aggregation by mercury compounds. *Throm. Res.* 1990; 57:795-801.

[28] Albert CM, Hennekens CH, O'Donnell CJ et al. Fish consumption and risk of sudden cardiac death. *JAMA* 1998; 279:23-28.

[29] Rinaldi S, Adembri C, Grechi S et al. Low-dose hydrocortisone during severe sepsis: effects on microalbuminuria. *Crit. Care Med.* 2006; 34:2334-2339.

[30] Tagawa H, Shimokawa H, Tagawa T et al. Long-term treatment with eicosapentaenoic acid augments both nitric oxide-mediated and non-nitric oxide-mediated endothelium-dependent forearm vasodilatation in patients with coronary artery disease. *J. Cardiovasc. Pharmacol.* 1999; 33:633-640.

[31] De Caterina R, Cybulsky MI, Clinton SK et al. The omega-3 fatty acid docosahexaenoate reduces cytokine-induced expression of proatherogenic and proinflammatory proteins in human endothelial cells. *Arterioscler. Thromb.* 1994; 14:1829-1836.

[32] Hughes DA, Pinder AC. n-3 polyunsaturated fatty acids inhibit the antigen-presenting function of human monocytes. *Am. J. Clin. Nutr.* 2000; 71:357S-360S.

[33] Caughey GE, Mantzioris E, Gibson RA et al. The effect on human tumor necrosis factor alpha and interleukin 1 beta production of diets enriched in n-3 fatty acids from vegetable oil or fish oil. *Am. J. Clin. Nutr.* 1996; 63:116-122.

[34] Johansen O, Seljeflot I, Hostmark AT et al. The effect of supplementation with omega-3 fatty acids on soluble markers of endothelial function in patients with coronary heart disease. *Arterioscler. Thromb. Vasc. Biol.* 1999; 19:1681-1686.

[35] Kang JX, Leaf A. Evidence that free polyunsaturated fatty acids modify Na+ channels by directly binding to the channel proteins. *Proc. Natl. Acad. Sci. U S A* 1996; 93:3542-3546.

[36] Clarke SD, Jump DB. Regulation of gene transcription by polyunsaturated fatty acids. *Prog. Lipid Res.* 1993; 32:139-149.

[37] Covas MI. Olive oil and the cardiovascular system. *Pharmacological Research* 2007; 55:175-186.

[38] Estruch R, Martinez-Gonzalez MA, Corella D et al. Effects of a Mediterranean-style diet on cardiovascular risk factors: a randomized trial. *Ann. Intern. Med.* 2006; 145:1-11.

[39] Leite MS, Pacheco P, Gomes RN et al. Mechanisms of increased survival after lipopolysaccharide-induced endotoxic shock in mice consuming olive oil-enriched diet. *Shock* 2005; 23:173-178.

[40] Paniagua JA, de la Sacristana AG, Sanchez E et al. A MUFA-rich diet improves posprandial glucose, lipid and GLP-1 responses in insulin-resistant subjects. *J. Am. Col.l Nutr.* 2007; 26:434-444.

[41] Perez-Martinez P, Lopez-Miranda J, Blanco-Colio L et al. The chronic intake of a Mediterranean diet enriched in virgin olive oil, decreases nuclear transcription factor kappaB activation in peripheral blood mononuclear cells from healthy men. *Atherosclerosis* 2007; 194:e141-e146.

[42] Nakamura N, Hamazaki T, Ohta M et al. Joint effects of HMG-CoA reductase inhibitors and eicosapentaenoic acids on serum lipid profile and plasma fatty acid concentrations in patients with hyperlipidemia. *Int. J. Clin. Lab. Res.* 1999; 29:22-25.

[43] Contacos C, Barter PJ, Sullivan DR. Effect of pravastatin and omega-3 fatty acids on plasma lipids and lipoproteins in patients with combined hyperlipidemia. *Arterioscler. Thromb* 1993; 13:1755-1762.

Conflict of Interest: The authors have no financial interest in any of the products mentioned in this paper.

In: Fish Consumption and Health
Editors: George P. Gagne and Richard H. Medrano
ISBN 978-1-60741-151-2

Chapter VII

Fish Consumption and Cancer: Summary of Evidence

Faina Linkov[1], Barbara Stadterman[1] and Emanuela Taioli * [1,2,3]
[1]University of Pittsburgh Cancer Institute, Department of Medicine, 5117 Centre Avenue, Pittsburgh, PA 15213, USA
[2]Department of Epidemiology, University of Pittsburgh Graduate School of Public Health, USA
[3]: SUNY Downstate School of Public Health
Graduate Program in Public Health at SUNY Downstate Medical Center

Abstract

Introduction

In the past few decades, there has been an enormous interest in quantifying the relationship between fish consumption and the risk of various malignancies. Previous research suggested that consumption of even relatively small amounts of fish is associated with a reduced risk of several cancers, especially of oral cavity and pharynx, esophagus, stomach, colon, rectum, and pancreas. Despite the growing body of literature in this area, studies have reported mixed results. The aim of this chapter is to provide a comprehensive summary of evidence that exists in the area of fish consumption and cancer by concentrating on four major cancers described in relation to fish consumption, including colorectal, breast, prostate, and lung cancers.

---

*To whom correspondence should be addressed: Emanuela Taioli, MD, PhD, SUNY Downstate School of Public Health; Graduate Program in Public Health at SUNY Downstate Medical Center; 450 Clarkson Avenue, Box 43, Brooklyn, New York 11203; Tel: (718)-804-7100
Fax: (718)-270-2533; email: emanuela.taioli@downstate.edu

Methods

Articles published from 1970 to 2008 were identified using the Pubmed search engine, and selected if they included results on the effect of dietary fish consumption on four major cancers, if they were cohort or case control studies, they were published in English or had expanded abstract available in English language. Case studies and studies with less than 10 participants were excluded. Two reviewers independently reviewed the articles using a structured abstraction form.

Results

Over 5,000 titles dealing with fish consumption and cancer have been identified through the initial search. Flagging the article titles that specifically dealt with the cancers of interest narrowed the search to approximately 50 studies in the area of biomarkers and 300 in the area of fish consumption and cancer. Seven studies in the area of biomarkers and fish consumption were evaluated in detail. These studies demonstrated that fish consumption plays an important role in inflammatory processes associated with cancer development.

Conclusions

Large number of cohort and case control studies from around the world provide interesting data about the efficacy of fish consumption for cancer prevention, however these data are still not conclusive. In particular, the effect of omega-3 fatty acid consumption on cancer incidence needs further investigation. There needs to be more well-designed longitudinal studies in this area to quantify the relationship between fish consumption and cancer.

Keywords: *fish consumption, tumor markers, risk factors, diet, lifestyle*

Background:

Cancer is one of the main leading causes of death in Western countries and around the world. Among the factors that contribute to the development of this disease, diet has a fundamental role, particularly in case of breast, colon-rectal, and prostate cancers[1]. This chapter will provide an overview of the knowledge in the area of fish consumption and cancer, specifically targeting the studies in the area of fish and cancer biomarkers.

Historical overview. Interest in the relationship between fish consumption and health came to the scientific arena in the early 1970's when two Danish physicians, Bang and Dyerberg, became aware that the mortality from heart disease among Greenland Eskimos was only about one-tenth that in Denmark and USA. These early findings sparked epidemiologic interest on the role of fish oil and fish consumption on health, as Dyerberg and Bang hypothesized that such a striking difference in cardiac mortality may be due to the different fats in the diet of Greenland Inuits in comparison to that of Danes and US citizens[2]. By early 1980's Bang, Dyerberg, and their colleagues concluded that the rarity of ischemic heart disease in Greenland Eskimos may partly be explained by the antithrombotic effect of the long-chained polyunsaturated fatty acids, especially eicosapentaenoic acid prevalent in diets rich in marine oils[3].

In the past several decades, many epidemiologic studies aimed to formalize and quantify the relationship between fish consumption, inflammation, cancer morbidity and mortality.

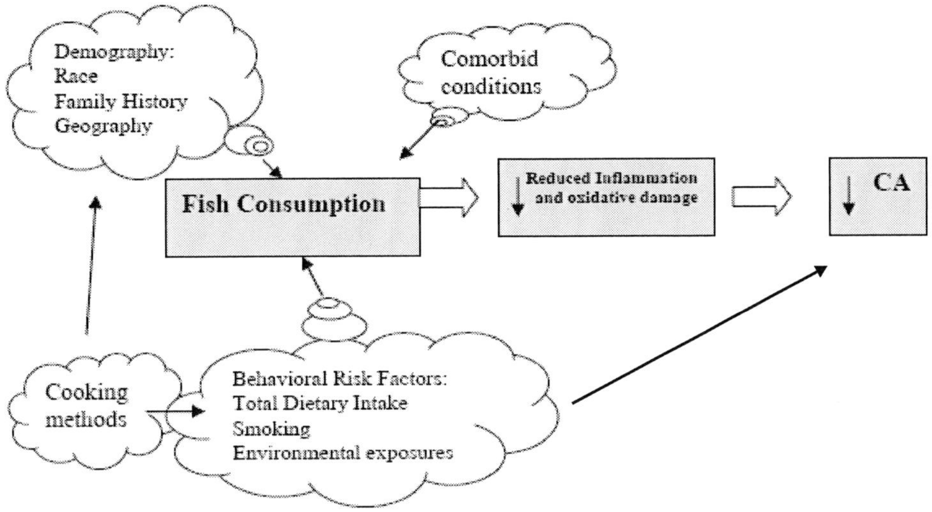

Figure 1. Fish: Protective effect on carcinogenesis development.

Habitual fish intake may attenuate the development of atherosclerosis and prevent ischemic heart disease through suppression of platelet activity, and by reducing platelet counts and serum triglyceride levels[4]. It has been hypothesized that fish consumption has a positive influence on many health conditions, including cardiovascular disease, cancer, and even on diseases associated with cognitive failure such as Alzheimer's. In the Zutphen Elderly Study[5], epidemiological data showed that among elderly men, fish consumers had less cognitive decline than did fish non-consumers. Figure 1 outlines the model for the positive influence of fish consumption on health developed by our group.

Recent Evidence on the Dangers of Fish Consumption

Controversial studies on fish consumption and cancer started to appear in the 1980's as a response to the new health recommendations encouraging fish consumption. There are several cancers that appear to be associated with fish consumption, especially nasopharyngeal carcinoma (NPC) and stomach cancer.

One of the earlier Japanese studies reported that the frequency of consumption of dried fish was hardly associated with deaths from cancer at all sites and from gastric cancer, but was significantly positively associated with deaths from cancer of the liver[6]. Epidemiological studies have associated the consumption of smoked fish and meat products with an increased risk of stomach cancer[7]. It is estimated that over 90% of young nasopharyngeal carcinoma cases (younger than 35) in Hong Kong Chinese can be attributed to consumption of Cantonese-style salted fish during childhood[8].

Exposure to salted fish was significantly associated with an increased risk of NPC. Four characteristics of exposure to salted fish independently contributed to the increased risk: (a) earlier age at first exposure, (b) increasing duration of consumption, (c) increasing frequency of consumption, and (d) cooking the fish by steaming it rather than frying, grilling, or boiling it[9].

Metals with major toxicological potential associated with the consumption of fishery products include antimony, arsenic, cadmium, chromium, lead, mercury, and nickel. Major organic contaminants that seem to pose sufficient risk include polychloricanted biphenils, dioxins, several chlorinated hydrocarbon insecticides, nitrosamines, chlorination products, and aquaculture related chemicals[10]. Because fish intake is a major source of exposure to mercury, the mercury content of fish may counteract the beneficial effects of its n-3 fatty acids[11].

Fish contains methylamines, which are precursors of N-nitrosamines. Nitrosamines can act as potent carcinogens in a wide variety of species[12] and therefore the National Academy of Science recommended that human exposure to N-Nitroso compouns should be avoided. Preliminary experimental data on Cantonese-style salted fish indicate that N-nitroso compounds may be involved in the carcinogenicity of this human food[13]. Figure 2 outlines the major pathways that lead to fish carcinogenicity. It is important to point out that cooking style plays a very important role in fish carcinogenicity, especially in the case of salted and smoked fish.

This chapter will focus on several major types of cancers investigated in relationship to fish consumption including breast, colorectal, prostate, and lung cancers.

Bioaccumulation and Biomagnificationof Toxins in Fish Tissues

Bioaccumulation and biomagnification of toxins in fish tissues is a serious concern. Mercury is one of the most widely investigated pollutants commonly found in fish, primarily the result of the deposition of industrial waste, from such sources as batteries, paints, pesticides and the burning of fossil fuels[14].

The methylation of mercury is a process by which it converts to methylmercury via aquatic micro-organisms. This most toxic form of mercury reaches even higher concentrations in the bodily tissues of both marine and saltwater fish through the processes known as bioaccumulation and biomagnifications. While fish ingest naturally-occurring environmental mercuric salts, the ratio of these elements of to that of highly toxic methylmercury, is reported to be approximately 1:9[15]. Persistent bioaccumulation and toxic substances (PBTS), such as methylmercury in fish and seafood may thus be associated with human body load of PBTS via consumption. Farming practices that utilize fishmeal increase the terrestrial food chain resulting in farm-animal accumulation of PBTS[16].

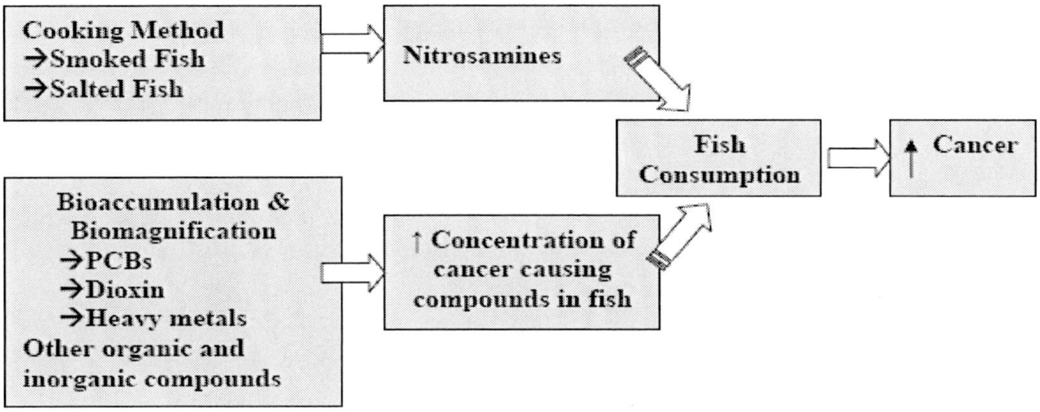

Figure 2. Fish as Carcinogen.

Several epidemiological studies have reliably linked dietary fish consumption with human exposure to mercury. A cohort study conducted between 1992 and 1995 by the European Prospective Investigation into Cancer (EPIC) in Gipuzkoa, a province of the Basque Country of Spain, examined blood mercury levels individuals as related to fish consumption via standardized dietary intake questionnaire method. This study showed a significant and direct association between fish consumption and methylmercury levels in red blood cells, an association that varied with daily intake[14].

A 2004 study conducted in New York City, in which blood levels of mercury were measured in a representative population of 1,811 residents, found similar association across racial and ethnic groups as a function of diet, with Asian populations, particularly those who emigrated from China having the highest concentrations of mercury in the blood. [17]. Studies conducted in Britain in 2006, also suggested a strong and direct correlation between blood, plasma and red cell concentrations indices of both mercury and selenium with the amount of oily fish consumed [18].

A recent Canadian investigation reported that metal concentrations in farmed and wild salmon from British Columbia, Canada, were relatively low and below human health consumption guidelines, with methylmercury in all salmon samples ranging from 0.03-0.1 microg/g wet wt. Canadian Health guidelines for safety 0.5 microg/g. Overall metal concentrations generally were higher in commercial salmon feed compared to farmed salmon, [22].

Blood mercury was likewise correlated with red-cell and plasma selenium that increased with age, possibly because of a chemical binding interaction that is biologically facilitated in humans. A French Polynesian study conducted in fishing a population from the South Pacific reported similarly high blood concentrations of mercury, selenium and omega-3 fatty acids, with quantities comparable to those reported among fishing populations from the Northern hemisphere[19]. While it has been suggested that selenium binds to mercury produces a decreased detriment to the body, the opposite has effect has been postulated as well[20, 21].

Bioaccumulation and biomagnification of dioxins (PCDD/Fs) and dioxin-like polychlorinated biphenyls in aquatic environment results in fishery products and by-products being an important vector to humans[23]. A 2007 study conducted on edible marine species (fish, cephalopod molluscs, crustaceans) from the Adriatic Sea, analyzed blood levels of heavy metals (Hg, Cd and Pb) as well as polychlorinated biphenyls (PCBs), and concluded that the risk to human health in terms of exposure to PCBs, cadmium and lead is negligible, while mercury exposure values due to consumption of certain fish species (albacore, rosefish and thornback ray) might be of concern[24].

The relationship between the ingestion of toxins via fish consumption as a risk factor for disease, particularly cancer, is a topic of continued research. A historical Japanese cohort study, conducted from 1973 to 1984 and from 1970 to 1981 respectively, examined cancer mortality among 1,351 patients with Minamata disease (MD) and 5,667 residents of coastal areas of Kagoshima. When adjusted for age, sex and follow-up period, no excess was noted for mortality from all causes, all cancers, and non-cancers combined, but the analysis of site-specific cancers showed a statistically significant decrease in mortality from stomach cancer and leukemia. [25].

USFDA Guidelines and Recommendations

The United States Food and Drug Administration (USFDA) has established guidelines for fish consumption in efforts to address health concerns.

The agency suggests that no amount of Shark, Swordfish, King Mackerel or Tilefish is safe, as they contain high levels of mercury. Up to 12 ounces per week of a variety of fish and shellfish that are lower in mercury can be safely consumed, such as shrimp, canned light tuna, salmon, pollock, and catfish. The most recent EPA advisory states that Albacore tuna contains higher levels of mercury than canned light tuna, and thus the FDA recommends consumption of only 6 ounces per week of albacore tuna. Fish caught in lakes, rivers, and coastal areas are subjected to the availability of information from local advisories; otherwise, consumers should eat only 6 ounces per week of it. A general recommendation is to consume a fish of different types and varied sources to decrease exposure risk.

Fish Consumption: Geographic Variation

Significant geographic variations in the trends of consumption, cooking, and the type of fish consumed (fatty vs. white) are present, which sometimes are difficult to interpret.

In Europe, intake of fish products is greater in northern parts than in the south. As a percentage of total fish consumption, intake of fatty fish is greater in the coastal areas of northern Europe (Denmark, Sweden) and in Germany than in central and southern Europe. Relatively few data are available for comparison of fish intake between countries, however researchers from EPIC study made on the first attempts to evaluate geographic variations between fish consumption in 10 different European countries. A six- to sevenfold variation in total fish consumption exists between the lowest consumption in Germany and the highest in Spain[59].

Globally, there exists significant variation in the consumption of fish as part of daily dietary intake. According to the Food and Agricultural Organization of the United Nations (FAO), twelve of the top fish consuming nations in the world consume more than 50 kg/capita/y, twice that of the United States which ranks at only 22.2 kg/capita/y. Maldives is the greatest consumer of fish, with a consumption rate of 154 kg/capita/y, followed by Iceland at 94 kg/capita/y, and Kiribati at 76.3 kg/capita/y. Other major fish consuming nations include Japan, French Polynesia, Norway and Seychelles, who consume between 60-70 kg/capita/y.

Fish Oil Supplements

Many people have turned to fish oil supplements to get adequate omega-3 fatty acids. Evidence from several studies has suggested that amounts of DHA and EPA in the form of fish or fish oil supplements lowers triglycerides, slows the buildup of atherosclerotic plaques lowers blood pressure slightly, as well as reduces the risk of death, heart attack, dangerous abnormal heart rhythms, and strokes in people with known heart disease. However, high doses may have harmful effects, such as an increased risk of bleeding. Although similar benefits are proposed for alpha-linolenic acid, scientific evidence is less compelling, and beneficial effects may be less pronounced [69]. Fish oil supplements have sometimes come under scrutiny in recent years. In early 2006, government agencies such as the Food Standards Agency in the UK and the Food Safety Authority of Ireland reported PCB levels that exceeded the strict new European maximum limits in several fish oil brands, which required temporary withdrawal of these brands.

Risks and benefits of fish and fish oil supplement consumptions need to be carefully evaluated before recommending any further changes to the diet in the US and around the world. While anyone can be harmed by high levels of mercury that can be found in fish like shark, some population groups are especially vulnerable to mercury and should consult their healthcare providers before consuming large amounts of fish that is known to contain mercury. The Centers for Disease Control and Prevention (CDC) reports that people most sensitive to mercury are pregnant and nursing women, children under the age of six (especially up to the age of three), people with impaired kidney function, and those with very sensitive immune responses to metals.

Future studies in this area should concentrate on examining the longitudinal changes in serum concentrations of cancer related biomarkers in relation to fish consumption. Geographic variation in fish consumption, type of fish consumed, and cooking styles, needs to be incorporated in these investigations. This data could be potentially investigated in relation to cancer incidence, treatment response, relapse, complications, and survival.

Methods

Search and selection of literature. A systematic literature search of publications on fish consumption and cancer development has been performed using Pubmed search engine. The following terms have been used in the database searches for studies on fish consumption and cancer development: fish, diet, risk factors, nutrition, fish. The criteria for inclusion of studies in the review were as follows: case control or cohort studies investigating the association between fish consumption and cancer development, and were published in English or had an expanded abstract in English.

In the area of biologic markers, all papers dealing with serum and plasma markers investigated in relation to fish consumption and cancer have been evaluated. The following terms have been used in the database searches for studies on endometrial cancer biomarkers: fish consumption, biomarkers, serum, plasma, inflammatory markers. The bibliographic lists of relevant publications were hand searched to identify additional useful articles. Articles were excluded if they had <10 cancer cases included in the analysis and if they were not full length research publications. Figure 3 outlines the article selection steps utilized in this study.

Data extraction. The data were extracted and analyzed by the authors of this review. Study size, study type, study population, methodology, and key findings were documented and reported in table 1. Our biomarker review did not include research articles exploring fish oil supplements, as those were evaluated in detail by MacLean and colleagues in a systematic review recently published in JAMA[26].

Results

The largest body of literature exists for prostate, breast, colorectal, and lung cancer in relation to fish consumption (between 30 and 90 articles per cancer). Limited literature is available for aerodigestive, endometrial, bladder, lymphoma, ovarian, pancreatic, skin, and stomach cancer. Basic developments in the fields of various cancers and fish consumptions were summarized in the next parts of this chapter. Table 1 summarizes findings in the area of cancer biomarkers.

Breast Cancer

Each year approximately one million women are newly diagnosed with breast cancer worldwide. In several studies in which the effect of fish consumption on cancer risk was investigated, meat and meat products were positively related to cancer risk, suggesting that cancer risks might be reduced more effectively when meat and meat products in meals are replaced by fish[27]. Essential fatty acids may modulate breast cancer risk by interacting with prostaglandins, which have immunosuppressive and platelet aggregative capabilities[28].

Existing studies also demonstrated that of the dietary components considered, percent calories from fish was the factor most strongly correlated with breast cancer rates after statistical adjustment for dietary fat intake[29] The results showed an inverse association between percent calories from fish and breast cancer rates that was consistent with a protective effect. This analysis confirmed the finding of others that dietary fat is strongly associated with international variation in breast cancer rates

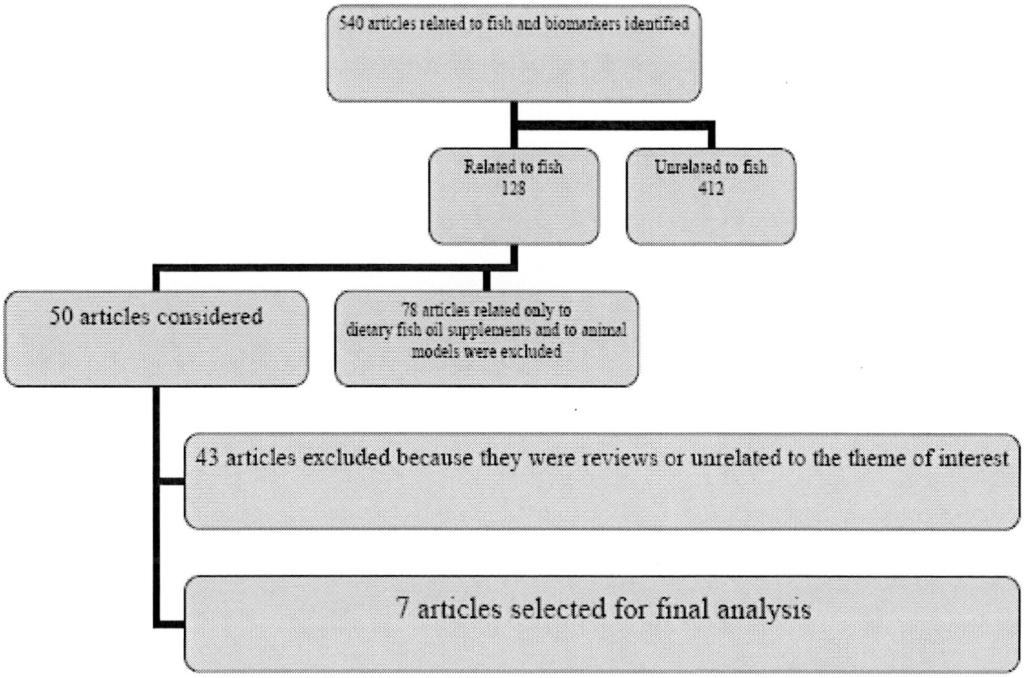

Figure 3. Article review flow for selecting literature on inflammation and fish consumption.

It also showed that of the dietary components considered, percent calories from fish was the factor most strongly correlated with breast cancer rates after statistical adjustment for dietary fat intake. This result is therefore in accord with animal experimental data and suggests that the omega-3 fatty acids contained in certain fish may protect against breast cancer[29].

Some in vitro and animal studies have suggested an inhibitory effect of marine n-3 fatty acids on breast cancer growth, but the results from epidemiological studies that have examined the association between fish consumption and breast cancer risk in humans are inconsistent. EPIC study provided no evidence for an association between fish intake and breast cancer risk[30]. Data from the Japan Collaborative Cohort (JACC) study suggested that there is a protective effects of the long-chain n-3 fatty acids that are abundant in fish[31]. Case control study carried out in Sweden demonstrated that high consumption of fish was weakly associated with reduced breast cancer risk, and the association was not statistically significant[32]. Mortality data for breast and colorectal cancer in 24 European countries were correlated with current fish and fish oil consumption and with consumption 10 and 23 years previously. There was no correlation at all between breast cancer mortality and fish or fish oil consumption at any time[33]. In Norway, compared with the reference group (wives of unskilled workers) the fishermen's wives had a decreased risk of breast cancer with a relative risk (RR) of 0.67 <95% confidence interval In = 0.47-0.94> adjusted for age and the number of children[34].

Colorectal Cancer

The age-adjusted incidence of colorectal cancer is higher in prosperous industrialized countries than elsewhere. Dietary factors may account for 75% of sporadic colorectal cancer in the west, with limited evidence suggesting protective effect of fish consumption on colorectal cancer development [35]. Chronic inflammation and colorectal cancer are closely linked. From a dietary perspective, fish oil containing (n-3) polyunsaturated fatty acids (PUFAs) has antiinflammatory properties, but for years the mechanism has remained obscure[36]. The European Prospective Investigation into Cancer and Nutrition (EPIC), which was specifically designed to investigate the relationship between diet and cancer reported that fish intake is negatively associated with colorectal cancer risk[37]. In cohort studies, fish consumption was shown to slightly reduce colorectal cancer risk[38]. Data from the Fukuoka Colorectal Cancer Study, a population-based case-control study, covering 782 cases and 793 controls, suggested that high intake of fish may decrease the risk, particularly of distal colon cancer[39]. There are substantial data from experimental studies in support of anticarcinogenic effects of fish/ n-3 polyunsaturated fatty acids in the colon. Several epidemiological studies have also provided evidence that fish/n-3 polyunsaturated fatty acids have anticarcinogenic effects in the colon, but not all data are consistent[40].

To clarify the association between intake of fish and the likelihood of developing colorectal cancer, researchers conducted a large sample size case-reference study with 928 cases of colon cancer, 622 of rectal cancer and 46886 cancer-free outpatient references aged 40-79 years. The data showed frequent raw/cooked fish intake to be associated with decreased odds ratio (OR) 0.68 with 95% confidence interval (CI) 0.47-0.99 for male colon cancer, especially for males aged over 60 years, smokers and frequent meat eaters. [41]. Another study demonstrated an inverse correlation with fish and fish oil consumption, when expressed as a proportion of total or animal fat, and this correlation was significant for both male and female colorectal cancer and for female breast cancer [42].

The role of fish in colorectal cancer development and progression is a little bit controversial, as Caygill and colleagues concluded that fish consumption is associated with protection against the later promotional stages of colorectal carcinogenesis, but not with the early initiation stages[33].

Fish consumption was shown to be protective for colorectal cancer in case control study conducted as a part of Melbourne Colorectal Cancer Study [43].

As primary prevention, a diet rich in vegetables, fruits, whole grain products, and legumes added by low-fat dairy products, fish, and poultry can be recommended[44].

Prostate Cancer

Consumption of fatty fish might reduce the risk of prostate cancer, the most frequent malignancy in males in the US. From 1930 to 1992, prostate cancer mortality rates in the

United States doubled and then declined somewhat until 2000[45]. In the majority of studies in this area, it was observed that high consumption of meat and dairy products has been linked to a greater risk. In contrast, frequent consumption of fatty fish and tomato products has been associated with a reduced risk[46].

Epidemiological studies suggest that fish consumption may be a protective factor against the development of prostate cancer. Mina et al. (2008) investigated the association of prostate cancer risk with fresh and preserved fish consumption among participants of a population-based case-control study (1,534 cases, 1,607 controls). Fish and poultry appear to be favorable diet indicators for prostate cancer[47]. Population based case control study in Canada suggested that consumption of preserved fish may reduce the risk of developing prostate cancer[48]. Frequent consumption of fatty fish and marine fatty acids appears to reduce the risk of prostate cancer, and this association is modified by genetic variation in the COX-2 gene[49].

Case-control study of diet and prostate cancer in Japan provided support to the hypothesis that the traditional Japanese diet, which is rich in soybean products and fish, might be protective against prostate cancer[50]. Marine fatty acids, particularly the long-chain eicosapentaenoic and docosahexaenoic acids, have been consistently shown to inhibit the proliferation of breast and prostate cancer cell lines in vitro and to reduce the risk and progression of these tumors in animal experiments[51].

Follow up of 47882 men participating in the Health Professionals Follow-up Study lead to the conclusion that men with high consumption of fish had a lower risk of prostate cancer, especially for metastatic cancer[52]. In Sweden, researchers studied the association between fish consumption and prostate cancer in a population-based prospective cohort of 6272 men. During 30 years of follow-up, men who ate no fish had a two-fold to three-fold higher frequency of prostate cancer than those who ate moderate or high amounts did[53]. Case control study in Poland demonstrated that the men who ate smoked or fried fish or liver at least once per week had almost half of the risk of prostate cancer of the men who ate those food rarely[54].

Gastric Cancer

The incidence of gastric cancer has greatly increased in the past 2-3 decades, however, the risk factors for the disease are still not clearly understood. An increased risk of stomach cancer was noted among those with high consumption of stewed foods such as soybean paste stew and hot pepper-soybean stew, broiled fish, and those who liked salty food[55].

Lung Cancer

N-3 polyunsaturated fatty acids in fish oil exhibit a variety of health benefits, and there is evidence that they can inhibit the development of human lung mucoepidermoid and other carcinomas. Researchers concluded that frequent fresh fish consumption, irrespective of the cooking method, may reduce the risk of lung cancer[56].

Esophageal Cancer

Evidence from animal studies indicates that various N-nitroso compounds are carcinogenic[57]. Previous studies have suggested that fermented fish sauce is related to an increased risk for nasopharyngeal, thyroid and gastric cancers and has suspicious carcinogenic and promoting effects in the laboratory[58].

Cancer and Biologic Markers: Importance of Evaluating the Effect of Fish Consumptions on Markers that Play an Important Role in Cancer Development.

A tumor biomarker is generally defined as an antigen or protein, which is secreted by the tumor itself or the surrounding tissues in response to the tumor and can be determined in the serum samples of the patient [60]. Biomarkers are cellular or soluble indicators of the physiological state and also of change during a disease process. Tumor biomarkers are comprised of factors that are produced either by tumor, stromal cells, or by surrounding tissues in response to tumor growth. Production of these factors plays a critical role in the interplay between several regulatory networks, such as immune system, angiogenic system, endocrine system, etc, that may affect tumor development resulting in either promotion or inhibition of tumor growth, progression, invasion and metastasis.

Despite the big role of biomarkers for cancer research especially in relation to cancer development, very few groups are investigating the longitudinal changes in biomarkers in same cohorts of patients. This review attempts to summarize the knowledge on inflammatory biomarkers involved in cancer development and fish consumption.

Table 1. Inflammatory markers involved in cancer development and progression and fish consumption

Reference	Study type/size	Biomarker name	Biomarker function	Findings
Tsitouras [61] 2008	Prospective diet intervention	CRP, IL-6	CRP-Acute phase protein associated with Inflammation IL-6-anti-inflammatory and pro-inflammatory cytokine	Serum C-reactive protein was significantly reduced and a trend towards lower IL-6 was noted with fish consumption.
Nanri [62] 2008	7802 subjects survey study	Hs-CRP	Acute phase protein associated with Inflammation (High sensitivity test for inflammatory marker has been	High intake of vegetables, fruit, soy products, and fish, was significantly and inversely related to hs-CRP concentrations in

Reference	Study type/size	Biomarker name	Biomarker function	Findings
			utilized)	both men and women.
Nettleton [63] 2006	5089 nondiabetic participants in the Multi-Ethnic Study of Atherosclerosis	Il-6	IL-6-anti-inflammatory and pro-inflammatory cytokine	Vegetables and fish diet was inversely related to IL-6
Niu [64] 2006	971 participants/cross sectional survey	CRP	Acute phase protein associated with Inflammation	Greater intake of n-3 PUFAs derived from marine products, as measured with a self-administered questionnaire, was independently related to a lower prevalence of high CRP.
Zampelas [65] 2006	ATTICA cross sectional survey of 1,514 men	CRP IL-6 TNF-alpha	TNF-alpha-cytokine involved in systemic inflammation SAA-protein that responds to inflammatory stimuli	Compared to non-fish consumers, those who consumed >300 g of fish per week had on average 33% lower CRP, 33% lower IL-6, 21% lower TNF-alpha, 28% lower SAA levels, and 4% lower WBC counts (all $p < 0.05$).
Lopez-Garcia[66] 2004	Cross-sectional study of 732 women from the Nurses' Health Study	CRP E-selectin	CRP-Acute phase protein associated with Inflammation E-selectin-cell adhesion molecule implicated in immune function	Higher intakes of fruit, vegetables, legumes, fish, poultry, and whole grains was inversely associated with plasma concentrations of CRP ($P = 0.02$) and E-selectin ($P = 0.001$) after adjustment for age,

Reference	Study type/size	Biomarker name	Biomarker function	Findings
				body mass index (BMI), physical activity, smoking status, and alcohol consumption.
Giovannucci [67] 2003	753 men/cross sectional survey	IGF-I	Protein hormone, stimulator of cell growth	The major sources of animal protein, including milk, fish, and poultry, but not red meat, as well as total vegetable protein, were associated with an increase in IGF-I levels.

Conclusion

Epidemiologic investigations on the association between fish consumption and cancer are rare. Existing studies in the area of fish consumption and cancer have many challenges including recall bias, confounding variables, presence of comorbid conditions, challenges associated with quantifying fish consumption and its effect. Fish consumption is an important modifiable factor that can influence health, especially the risk of cancer development. More studies need to be developed where the benefits of fish consumption are carefully evaluated against the risks. There is a need to harness the knowledge of the mechanisms of cancer prevention related to dietary fish, in order to develop new preventive therapies and strategies. In addition, there is a need for more studies investigating the relationship between fish consumption and other risk factors playing a role in cancer development. The fundamental principles of optimal health and optimal ageing are abstaining from smoking, modest alcohol consumption, regular physical exercise and a diet rich in fish and plants and low in condensed calories, sugar and dairy products[68], however due to the plethora of risk factors associated with cancer development it is hard to quantify the exact relationship between cancer development and fish consumption.

References

[1] Granados, S., et al., Dietary lipids and cancer. *Nutr Hosp*, 2006. 21 Suppl 2: p. 42-52, 44-54.

[2] Leaf, A. The Beneficial Effects of Long-Chain, Polyunsaturated n-3 Fish Oil Fatty Acids on the Cardiovascular System. 2006 [cited 2008 March 5]; Available from: http://www.pitt.edu/~super1/lecture/lec23181/index.htm.

[3] Bang, H.O., J. Dyerberg, and H.M. Sinclair, The composition of the Eskimo food in north western Greenland. *Am J Clin Nutr*, 1980. 33(12): p. 2657-61.

[4] Imano, H., et al., [The effects of fish supplementation of platelet function, count and metabolism in healthy Japanese]. *Nippon Eiseigaku Zasshi*, 1999. 53(4): p. 601-10.

[5] Connor, W.E. and S.L. Connor, The importance of fish and docosahexaenoic acid in Alzheimer disease. *Am J Clin Nutr*, 2007. 85(4): p. 929-30.

[6] Ikeda, M., et al., A cohort study on the possible association between broiled fish intake and cancer. *Gann,* 1983. 74(5): p. 640-8.

[7] Ohshima, H., et al., Formation of direct-acting genotoxic substances in nitrosated smoked fish and meat products: identification of simple phenolic precursors and phenyldiazonium ions as reactive products. *Food Chem Toxicol*, 1989. 27(3): p. 193-203.

[8] Yu, M.C., et al., Cantonese-style salted fish as a cause of nasopharyngeal carcinoma: report of a case-control study in Hong Kong. *Cancer Res, 1986*. 46(2): p. 956-61.

[9] Ning, J.P., et al., Consumption of salted fish and other risk factors for nasopharyngeal carcinoma (NPC) in Tianjin, a low-risk region for NPC in the People's Republic of China. *J Natl Cancer Inst*, 1990. 82(4): p. 291-6.

[10] Ahmed, F.E., et al., Human health risks due to consumption of chemically contaminated fishery products. *Environ Health Perspect,* 1993. 101 Suppl 3: p. 297-302.

[11] Guallar, E., et al., Mercury, fish oils, and the risk of myocardial infarction. *N Engl J Med*, 2002. 347(22): p. 1747-54.

[12] Zeisel, S.H. and K.A. DaCosta, Increase in human exposure to methylamine precursors of N-nitrosamines after eating fish. *Cancer Res,* 1986. 46(12 Pt 1): p. 6136-8.

[13] Yu, M.C. and B.E. Henderson, Intake of Cantonese-style salted fish as a cause of nasopharyngeal carcinoma. *IARC Sci Publ*, 1987(84): p. 547-9.

[14] Sanzo, J.M., et al., Estimation and validation of mercury intake associated with fish consumption in an EPIC cohort of Spain. *Public Health Nutr*, 2001. 4(5): p. 981-8.

[15] Langworth, S., et al., Biological monitoring of environmental and occupational exposure to mercury. *Int Arch Occup Environ Health*, 1991. 63(3): p. 161-7.

[16] Dorea, J.G., Persistent, bioaccumulative and toxic substances in fish: Human health considerations. *Sci Total Environ*, 2008.

[17] McKelvey, W., et al., A biomonitoring study of lead, cadmium, and mercury in the blood of New York City adults. *Environ Health Perspect*, 2007. 115(10): p. 1435-41.

[18] Bates CJ, P.A., Birch MC, Delves HT. , Dependence of blood indices of selenium and mercury on estimated fish intake in a national survey of British adults. *Public Health Nutr*. 2007 May;10(5):508-17. *Public Health Nutr*, 2007. 10(5): p. 508-17.

[19] Dewailly, E., L. Chateau-Degat, and E. Suhas, Fish consumption and health in French Polynesia. Asia Pac J *Clin Nutr*, 2008. 17(1): p. 86-93.

[20] Hansen, J.C., Has selenium a beneficial role in human exposure to inorganic mercury? *Med Hypotheses*, 1988. 25(1): p. 45-53.

[21] Chen, C., et al., Accumulation of mercury, selenium and their binding proteins in porcine kidney and liver from mercury-exposed areas with the investigation of their redox responses. *Sci Total Environ*, 2006. 366(2-3): p. 627-37.

[22] Kelly, B.C., et al., Mercury and other trace elements in farmed and wild salmon from British Columbia, Canada. *Environ Toxicol Chem*, 2008. 27(6): p. 1361-70.

[23] Blanco, S.L., et al., Dietary uptake of dioxins (PCDD/PCDFs) and dioxin-like PCBs in Spanish aquacultured turbot (Psetta maxima). *Food Addit Contam*, 2007. 24(4): p. 421-8.

[24] Storelli, M.M., Potential human health risks from metals (Hg, Cd, and Pb) and polychlorinated biphenyls (PCBs) via seafood consumption: Estimation of target hazard quotients (THQs) and toxic equivalents (TEQs). *Food Chem Toxicol*, 2008. 46(8): p. 2782-8.

[25] Kinjo, Y., et al., Cancer mortality in Minamata disease patients exposed to methylmercury through fish diet. *J Epidemiol*, 1996. 6(3): p. 134-8.

[26] MacLean, C.H., et al., Effects of omega-3 fatty acids on cancer risk: a systematic review. *Jama,* 2006. 295(4): p. 403-15.

[27] de Deckere, E.A., Possible beneficial effect of fish and fish n-3 polyunsaturated fatty acids in breast and colorectal cancer. *Eur J Cancer Prev*, 1999. 8(3): p. 213-21.

[28] Godley, P.A., Essential fatty acid consumption and risk of breast cancer. *Breast Cancer Res Treat*, 1995. 35(1): p. 91-5.

[29] Kaizer, L., et al., Fish consumption and breast cancer risk: an ecological study. *Nutr Cancer*, 1989. 12(1): p. 61-8.

[30] Engeset, D., et al., Fish consumption and breast cancer risk. The European Prospective Investigation into Cancer and Nutrition (EPIC). Int J Cancer, 2006. 119(1): p. 175-82.

[31] Wakai, K., et al., Dietary intakes of fat and fatty acids and risk of breast cancer: a prospective study in Japan. *Cancer Sci*, 2005. 96(9): p. 590-9.

[32] Terry, P., et al., Fish consumption and breast cancer risk. *Nutr Cancer*, 2002. 44(1): p. 1-6.

[33] Caygill, C.P. and M.J. Hill, Fish, n-3 fatty acids and human colorectal and breast cancer mortality. *Eur J Cancer Prev*, 1995. 4(4): p. 329-32.

[34] Lund, E. and K.H. Bonaa, Reduced breast cancer mortality among fishermen's wives in Norway. Cancer Causes Control, 1993. 4(3): p. 283-7.

[35] Johnson, I.T. and E.K. Lund, Review article: nutrition, obesity and colorectal cancer. Aliment Pharmacol Ther, 2007. 26(2): p. 161-81.

[36] Chapkin, R.S., et al., Immunomodulatory effects of (n-3) fatty acids: putative link to inflammation and colon cancer. *J Nutr*, 2007. 137(1 Suppl): p. 200S-204S.

[37] Gonzalez, C.A. and E. Riboli, Diet and cancer prevention: where we are, where we are going. *Nutr Cancer*, 2006. 56(2): p. 225-31.

[38] Geelen, A., et al., Fish consumption, n-3 fatty acids, and colorectal cancer: a meta-analysis of prospective cohort studies. *Am J Epidemiol*, 2007. 166(10): p. 1116-25.

[39] Kimura, Y., et al., Meat, fish and fat intake in relation to subsite-specific risk of colorectal cancer: The Fukuoka Colorectal Cancer Study. *Cancer Sci*, 2007. 98(4): p. 590-7.

[40] Kimura, Y., [Fish, n-3 polyunsaturated fatty acid and colorectal cancer prevention: a review of experimental and epidemiological studies]. *Nippon Koshu Eisei Zasshi*, 2006. 53(10): p. 735-48.

[41] Yang, C.X., et al., Fish consumption and colorectal cancer: a case-reference study in Japan. *Eur J Cancer Prev*, 2003. 12(2): p. 109-15.

[42] Caygill, C.P., A. Charlett, and M.J. Hill, Fat, fish, fish oil and cancer. *Br J Cancer*, 1996. 74(1): p. 159-64.

[43] Kune, S., G.A. Kune, and L.F. Watson, Case-control study of dietary etiological factors: the Melbourne Colorectal Cancer Study. *Nutr Cancer*, 1987. 9(1): p. 21-42.

[44] Strohle, A., W. Maike, and A. Hahn, [Nutrition and colorectal cancer]. *Med Monatsschr Pharm*, 2007. 30(1): p. 25-32.

[45] Colli, J.L. and A. Colli, Comparisons of prostate cancer mortality rates with dietary practices in the United States. Urol Oncol, 2005. 23(6): p. 390-8.

[46] Wolk, A., Diet, lifestyle and risk of prostate cancer. *Acta Oncol*, 2005. 44(3): p. 277-81.

[47] Hu, J., et al., Meat and fish consumption and cancer in Canada. *Nutr Cancer*, 2008. 60(3): p. 313-24.

[48] Mina, K., L. Fritschi, and K.C. Johnson, An inverse association between preserved fish and prostate cancer: results from a population-based case-control study in Canada. *Nutr Cancer*, 2008. 60(2): p. 222-6.

[49] Hedelin, M., et al., Association of frequent consumption of fatty fish with prostate cancer risk is modified by COX-2 polymorphism. *Int J Cancer*, 2007. 120(2): p. 398-405.

[50] Sonoda, T., et al., A case-control study of diet and prostate cancer in Japan: possible protective effect of traditional Japanese diet. *Cancer Sci*, 2004. 95(3): p. 238-42.

[51] Terry, P.D., T.E. Rohan, and A. Wolk, Intakes of fish and marine fatty acids and the risks of cancers of the breast and prostate and of other hormone-related cancers: a review of the epidemiologic evidence. *Am J Clin Nutr*, 2003. 77(3): p. 532-43.

[52] Augustsson, K., et al., A prospective study of intake of fish and marine fatty acids and prostate cancer. *Cancer Epidemiol Biomarkers Prev*, 2003. 12(1): p. 64-7.

[53] Terry, P., et al., Fatty fish consumption and risk of prostate cancer. *Lancet*, 2001. 357(9270): p. 1764-6.

[54] Pawlega, J., J. Rachtan, and T. Dyba, Dietary factors and risk of prostate cancer in Poland. Results of case-control study. *Neoplasma*, 1996. 43(1): p. 61-3.

[55] Lee, J.K., et al., Dietary factors and stomach cancer: a case-control study in Korea. *Int J Epidemiol*, 1995. 24(1): p. 33-41.

[56] Takezaki, T., et al., Diet and lung cancer risk from a 14-year population-based prospective study in Japan: with special reference to fish consumption. *Nutr Cancer*, 2003. 45(2): p. 160-7.

[57] Rogers, M.A., et al., Consumption of nitrate, nitrite, and nitrosodimethylamine and the risk of upper aerodigestive tract cancer. *Cancer Epidemiol Biomarkers Prev*, 1995. 4(1): p. 29-36.

[58] Ke, L., P. Yu, and Z.X. Zhang, Novel epidemiologic evidence for the association between fermented fish sauce and esophageal cancer in South China. *Int J Cancer*, 2002. 99(3): p. 424-6.

[59] Welch, A.A., et al., Variability of fish consumption within the 10 European countries participating in the European Investigation into Cancer and Nutrition (EPIC) study. *Public Health Nutr*, 2002. 5(6B): p. 1273-85.

[60] Canney, P.A., et al., Ovarian cancer antigen CA125: a prospective clinical assessment of its role as a tumour marker. Br J Cancer, 1984. 50(6): p. 765-9.

[61] Tsitouras, P.D., et al., High omega-3 fat intake improves insulin sensitivity and reduces CRP and IL6, but does not affect other endocrine axes in healthy older adults. *Horm Metab Res*, 2008. 40(3): p. 199-205.

[62] Nanri A, Y.D., Yamaji T, Mizoue T, Takayanagi R, Kono S., Dietary patterns and C-reactive protein in Japanese men and women. 2008. 87(5): p. 1486-96.

[63] Nettleton, J.A., et al., Dietary patterns are associated with biochemical markers of inflammation and endothelial activation in the Multi-Ethnic Study of Atherosclerosis (MESA). *Am J Clin Nutr*, 2006. 83(6): p. 1369-79.

[64] Niu, K., et al., Dietary long-chain n-3 fatty acids of marine origin and serum C-reactive protein concentrations are associated in a population with a diet rich in marine products. *Am J Clin Nutr*, 2006. 84(1): p. 223-9.

[65] Zampelas, A., et al., Fish consumption among healthy adults is associated with decreased levels of inflammatory markers related to cardiovascular disease: the ATTICA study. *J Am Coll Cardiol*, 2005. 46(1): p. 120-4.

[66] Lopez-Garcia, E., et al., Major dietary patterns are related to plasma concentrations of markers of inflammation and endothelial dysfunction. *Am J Clin Nutr*, 2004. 80(4): p. 1029-35.

[67] Giovannucci, E., et al., Nutritional predictors of insulin-like growth factor I and their relationships to cancer in men. *Cancer Epidemiol Biomarkers Prev*, 2003. 12(2): p. 84-9.

[68] Bengmark, S., Impact of nutrition on ageing and disease. Curr Opin Clin Nutr Metab Care, 2006. 9(1): p. 2-7.

[69] MedlinePlus. Omega-3 fatty acids, fish oil, alpha-linolenic acid. 2008 [cited 2008 July 30]; Available from: http://www.nlm.nih.gov/medlineplus/druginfo/natural/patient-fishoil.html.

In: Fish Consumption and Health
Editors: George P. Gagne and Richard H. Medrano
ISBN 978-1-60741-151-2

Chapter VIII

Federal and State Advisories on Mercury in Seafood: An Assessment of Women's Fish Consumption Behaviors and Advisory Awareness

***Carole B. Rudra*[1,2] *and Michelle A. Williams*[2,3]**
[1]Department of Social and Preventive Medicine, University at Buffalo, State University of New York. Buffalo, NY, USA
[2] Center for Perinatal Studies, Swedish Medical Center, Seattle, WA, USA
[3] Department of Epidemiology, School of Public Health and Community Medicine, University of Washington, Seattle, WA, USA

Abstract

Fish consumption confers nutritional benefits and potential contamination risks. The issue is of particular concern to reproductive-aged women and their infants. Although federal and local fish consumption advisories are controversial and appear to cause confusion, their behavioral and biological impacts are not well understood. We review available literature regarding correlates of fish consumption, the extent of advisory awareness among Americans, and the behavioral impacts of advisory awareness. We estimate advisory compliance and its sociodemographic and behavioral correlates among pregnant participants of a western Washington cohort study. Finally, we identify gaps in current knowledge and suggest areas for future research. Many studies show that maternal fish consumption directly relates to maternal and fetal mercury exposures. Far fewer describe women's knowledge and behaviors regarding the risks of fetal mercury exposures. Among 2,383 pregnant women in western Washington, lack of periconceptional fish consumption was associated with maternal age <20 years and both leanness and overweight before pregnancy. Consumption above the federal advisory guidelines was associated with the age group ≥ 35 years, with smoking before pregnancy, post-high school education, and nulliparity. Until we can better understand the behavioral impacts of federal and local advisories, we will be unable to improve the ways in which we communicate the benefits and risks of fish consumption to reproductive-aged women.

Introduction

Weighing the benefits and risks of fish consumption and communicating them to reproductive-aged women presents an ongoing public health challenge. Fish is an excellent source of protein, vitamin D, preformed long-chain polyunsaturated omega-3 fatty acids (PUFAs), and other nutrients. Maternal fish consumption may improve infant vision and cognition in infancy and childhood [1-5]. However, low-level prenatal methylmercury exposures have been linked to decreased cognition in infancy; language, attention, and memory deficits in childhood; and increased blood pressure and heart rate variability in childhood [6-9]. While many of these beneficial and adverse outcomes are subtle, their importance is great when considered in context of the population's exposure burden.

Nearly 6% of reproductive-aged American women have blood mercury concentrations above the level considered to be without appreciable harm to themselves or their infants [10]. Federal, state, and local governments advise pregnant and reproductive-aged women to limit their consumption of fish, the most common source of methylmercury exposure [11]. (Throughout this chapter, we use the term 'fish' to refer to freshwater and ocean fish, shellfish, and mollusks). Despite much controversy surrounding these advisories there is little information regarding the target population's awareness and responsiveness to them [12-14]. Here we will describe how fish consumption is related to maternal characteristics and mercury concentrations in maternal and fetal tissues. We will summarize literature regarding advisory awareness and its behavioral impacts. We will also report an original examination of fish consumption behaviors in pregnant women. Finally, we will identify gaps in the current research and propose approaches to fill them.

Fish Consumption: Risks and Benefits

Mercury is a persistent chemical arising from natural and anthropogenic sources [15]. Metallic, inorganic (ionic), ethyl-, and methylmercury are most important with regard to human toxicology. In humans, the primary source of metallic mercury exposure is amalgam dental fillings. Inorganic mercury, historically used in medicines and cosmetics, is now uncommon. Ethylmercury was used as a vaccine preservative in the U.S. until 1999 [15].

Upon entering water bodies after atmospheric deposition, mercury is converted to methylmercury by aquatic biota and bioaccumulates in aquatic food webs. Methylmercury is most highly concentrated in long-living predatory fish [16]. In humans, about 95% of methylmercury in consumed fish is absorbed in the gastrointestinal tract [15]. It is then nearly completely absorbed into the blood and is distributed to all tissues, including the placenta, within 30-40 hours [16, 17]. Adult and fetal methylmercury exposures are typically quantified by measuring methylmercury concentrations in blood, scalp hair, and cord blood [17].

Methylmercury concentrations vary widely between and within fish species [12, 18]. In an FDA grocery survey, concentrations ranged from undetectable to 1.4 ppm across species [19]. Concentrations in canned tuna, which comprises 20-25% of the fish consumed in the U.S, are particularly variable and may be increasing over time [20].

Health effects of methylmercury. The developmental impacts of methylmercury became apparent after episodes of high exposures in Japan and Iraq. The resulting neurological damages were much more severe than the effects in adults [15]. Neurological effects of *in utero* exposures at lower doses have subsequently been examined in several cohorts, as reviewed by Spurgeon [21]. The results of these studies have been inconsistent. In brief, two studies in New Zealand and the Faroe Islands provided evidence of neurodevelopmental effects, while another in the Seychelles Islands showed essentially no association [22-24]. In a later Polish cohort, fish consumption during pregnancy was inversely associated with indices of infant mental and psychomotor development at 12 months of age, but these associations were no longer apparent at 24 and 36 months of age [25].

Potential explanations for the inconsistencies across studies include differences in mercury biomarkers, neurodevelopmental measures, and influences of the consumption of marine mammals versus ocean fish [26-28]. Some experts note that the variation in these study findings is less inconsistent than differences in statistical significance would suggest, because the slopes of the associations are almost identical across studies [29]. Furthermore, analyses of neurological tests administered to Seychelles Islands cohort members at 9 years of age suggest that delayed adverse effects of higher exposures might be emerging as the children mature [26]. Damage from *in utero* exposure may manifest itself in higher-order cognitive functions that develop in late childhood and early adolescence [26].

Health effects of fish consumption. While the health effects of low-level *in utero* methylmercury exposures continue to be examined, evidence of the benefits of maternal fish consumption grows. The benefits likely arise mostly via PUFAs that are present in high concentrations in some species. Aggregated evidence from randomized controlled trials suggests that maternal dietary and infant formula PUFA supplementation can increase child intelligence quotient [1]. In one randomized clinical trial, fish oil supplementation reduced preterm delivery rates in high-risk women with low or moderate fish consumption during pregnancy, though not in women who reported high consumption [30]. PUFA supplementation and fish consumption during pregnancy have also been associated with longer gestation and higher birthweight in observational studies, though not consistently [31-33]. Given substantial evidence of PUFAs' cardiovascular benefits in non-pregnant populations, the American Heart Association (AHA) recommends eating fish, particularly fatty species, at least twice a week [34].

Reference dose and fish advisories. Based on data from the Iraq methylmercury exposure episode and prospective cohorts, the United States Environmental Protection Agency (EPA) defined a methylmercury reference dose as 0.1 μg/kg body weight/day [15]. This amount, consumed daily over a lifetime, is assumed to be without adverse health effects to humans, including fetuses. It is estimated to correspond with 5.8 μg/L methylmercury concentration in adult and cord blood and 1.2 ppm in hair [35].

The EPA and United States Food and Drug Administration (FDA) issued fish consumption advisories targeted to pregnant women in 2001 and again in 2004 [18]. The current advisory recommends three behaviors to women who are or might become pregnant and nursing mothers: (1) avoidance of shark, swordfish, king mackerel, and tilefish; (2) consumption of ≤12 oz./week of a variety of fish and shellfish that are lower in mercury, with ≤6 oz./week of albacore tuna; (3) checking local advisories about safety of locally caught fish, when available, or local fish consumption of ≤6 oz./week when local advisories are not available [18]. As of 2006, nearly 4,000 local fish advisories have also been issued by 44 states, one territory, and two tribes [36]. About 80% of these address mercury, while the remainder address other contaminants such as chlordane and dioxins.

Extent of exposure. At least a small proportion of American women are exposed to potentially harmful amounts of mercury. The EPA estimates that 1-3% of women aged 15-44 years eat enough fish to accumulate potentially harmful concentrations [16]. Using data from a seven-day recall survey of 1,000 New Jersey residents, Stern et al. estimated that 21-30% of women aged 18-40 exceeded the consumption reference dose [37]. The most representative U.S. data are from 16-49 year old women participating in the National Health and Nutrition Examination Survey (NHANES) from 1999 to 2002. Of 1,577 participants, 5.7% (95% confidence interval [CI], 4.0-8.0%) had blood total mercury concentrations exceeding the reference dose of 5.8 µg/L [10]. The median concentration was 0.26 µg/L. Based on this survey, Mahaffey et al. estimated that 300,000 newborns each year may be at risk for neurodevelopmental effects [38].

Controversy and confusion. The apparent contradiction between advisories and the AHA recommendation reflects the simultaneous putative benefits and risks of fish consumption. This incongruity has caused controversy and, perhaps, confusion among the public and health professionals [12, 14]. Researchers have argued against the validity of the EPA reference dose [13, 40, 41]. Others have criticized various aspects of the advisories: the lack of specificity with regard to body weight and portion size, the insufficient summary of the benefits of fish, and the need for more detail to help women identify low-mercury species [13, 40-42].

The Institute of Medicine (IOM) recently established a committee to examine evidence regarding the nutritional benefits and contamination risks [2]. The committee recommended that advice provided to women who are or may become pregnant should stay within limits prescribed by federal and state advisories. However, they also noted that advice from federal and private organizations has been fragmented and sometimes conflicting. In many instances, they concluded, the impact of dietary guidance is not well measured or understood. Recognizing that consumers do not always change behaviors after receiving dietary advice, the committee noted that increased understanding of the influences of consumer choice is necessary for advisory design.

A Review of Relevant Literature

We reviewed literature regarding fish consumption, methylmercury exposure, and advisory awareness. We searched Medline to identify English-language publications through

June 2008 [43]. Search terms included mercury, methylmercury, fish, seafood, diet, pregnancy, maternal, infant, fetal, advisory, and reproduction. We also searched the references of the publications we identified for additional relevant literature.

Fish consumption and mercury exposure. Several studies have assessed the relation between fish consumption and mercury concentrations in maternal and infant tissues [3, 25, 35, 44-49]. A few others have examined mercury biomarkers in reproductive-aged women [38, 50, 51]. These are summarized in Table 1. In nine of ten studies of pregnant women, fish consumption before or during pregnancy was associated with hair, maternal blood, or cord blood mercury [3, 25, 35, 44-49]. In contrast, Stern et al. found no association between hair mercury and fish consumption in the first and second trimesters [48]. One study did not directly assess the association but noted that both fish consumption and maternal blood methylmercury decreased through pregnancy [49]. Three others examined associations with individual species or types of fish. Björnberg et al. found that deep-frozen fish consumption was strongly associated with maternal hair and infant cord blood mercury [35]. Oken et al. examined four groups (canned tuna, shellfish, dark meat, and white meat fish) and found that all were moderately correlated with maternal hair mercury [3]. Ramón saw associations with large oily fish, lean fish, and mixed fresh fish, but not tuna, sardines, or smoked fish [52].

Two of these studies presented coefficients of association. Oken et al. reported that a one serving/week increase during the second trimester (serving size unspecified) was associated with a 0.17 ppm increase in hair mercury among Massachusetts residents (95% CI, 0.10-0.24 ppm) [3]. This increase comprises 14% of the EPA hair reference dose. Using NHANES data, Mahaffey et al. estimated that each 1 g serving/kg body weight in the past 30 days was associated with a 0.25 μg/L increase in blood mercury ($p<0.0001$) [38]. The analogous increase for a 65 kg woman consuming a weekly 6 oz. serving would be 2.8 μg/L, comprising 48% of the EPA blood reference dose.

This body of research considered populations that were diverse in geographic location, race/ethnicity, and fish consumption behaviors. Because investigators chose several exposure assessment windows, mercury biomarkers, and estimates of association, direct comparisons of their findings are difficult. However, the results strongly suggest associations between fish consumption and both maternal and fetal mercury exposures.

Fish consumption correlates. There has been little research on the sociodemographic and behavioral characteristics related to fish consumption in young women. Hightower et al. analyzed data from 140 female NHANES participants (1999-2002) aged 16-49 years who identified themselves as "other" race/ethnicity (Asian, Pacific Islander, Native American, or multiracial) [53]. The authors compared fish consumption (meals in the past 30 days) and blood mercury concentrations within this group to values within other racial/ethnic groups. This group reported more meals in the previous 30 days (mean, 8.0; 95% CI, 4.0-12.0) than groups classified as Mexican American, Non-Hispanic white, Non-Hispanic black, and other Hispanic (mean, 2.7-4.9 servings; see Figure 1). Similarly, the prevalence of reference dose exceedance was higher in this group (16.6%; 95% CI, 8.4-24.8%) than in others (prevalence, 1.7-5.8%, see Figure 1).

Table 1. Published studies of the association between fish consumption and biological mercury concentrations in pregnant or reproductive-aged women or their infants

Reference	Population	Fish consumption	Relation with biological concentrations
Cordier et al. 1998 [50]	109 pregnant women, 255 other adults, 136 children, French Guiana, 1994	99% ate fish; 22% ate ≥5 meals/wk.	Total, freshwater, and ocean fish meals/wk were correlated with hair T-Hg. Spearman ρ were 0.50, 0.34, and 0.62, respectively (all S).
Vahter et al. 2000a [49]	112 pregnant Swedish women, 1994-96	13% ate >0.23 meals/wk (freshwater) before pregnancy, 4% during pregnancy.	Blood MeHg decreased throughout pregnancy (S); authors hypothesized that this was due to decreased consumption.
Muckle et al. 2001[46]	135 pregnant Inuit women in Quebec, 1995-98	33% ate <1 meal/wk; mean was 3.3 meals/wk.	Consumption was not correlated with blood MeHg near term ($\rho = -0.09$, NS), but was with hair MeHg post-delivery ($\rho = 0.20$, S).
Stern et al. 2001 [48]a	189 pregnant NJ women, 1995-97	92% ate fish; mean was 1.6 meals/wk.	Hair T-Hg in trimester 1-2 and consumption were not correlated.
Björnberg et al. 2003 [35] b	123 pregnant Swedish women, 1996-99	99% ate any fish before pregnancy; freshwater fish averaged 0.54 oz/wk before, 0.12 oz/wk during pregnancy.	Hair T-Hg and cord MeHg were associated with total consumption ($\rho = 0.41$ and 0.46, respectively) but not freshwater fish consumption.
Morrissette et al. 2004 [45]	159 pregnant Quebecois women, date unspecified	Means were 0.83 meals/wk before pregnancy, 0.69 meals/wk during pregnancy.	Dose-response (S) in hair T-Hg over meals/wk. Cord blood organic Hg was correlated with meals/wk before ($\rho = 0.46$) and during pregnancy ($\rho = 0.45$) (both S).
Mahaffey et al. 2004 [38]	1,709 US female 16-49 year olds,1999-2000	14% ate fish or shellfish in past 24 hours.	Estimated MeHg from fish was correlated with blood MeHg ($\rho = 0.5$, S).

Table 1. (Continued)

Reference	Population	Fish consumption	Relation with biological concentrations
Oken et al. 2005 [3]	135 pregnant women in eastern MA, 1999-2003	Mean was 1.2 servings/wk in 2nd trimester.	Each serving/wk associated with 0.17 ppm increased Hair T-Hg (S). Tuna, white fish, dark fish, and shell fish were moderately correlated with hair T-Hg ($0.23 \leq \rho \leq 0.43$)
Sato et al. 2006 [47]	184 Hawaiian pregnant women, 2004-05	94% ate fish during pregnancy; 32% ate >1 meal/wk.	Fish in last month of pregnancy was associated with cord Hg (S).
Gibičar et al. 2006 [51]	291 Greek women pregnant or with child ≤4 years, date unspecified	90% ate fresh local fish; 55% ate >1 meal/wk.	Dose-responses were reported between meals/week and hair T-Hg and Me-Hg (S).
Kim et al. 2006 [44]	63 pregnant Korean women, 2004	89% ate fish in early pregnancy; 52% ate ≥0.9 serving/wk.	Median blood T-Hg was higher in those who ate ≥2.8 servings/wk. vs. 1 or none (S).
Jedrychowki et al. 2007b [25]	374 infants born to Polish women aged 18-35 years, 2001-2003	Mean 4.9 oz/wk in 1st and 2nd trimesters; 6.2 oz/wk in 3rd.	Cord T-Hg correlated with median consumption in 1st 2 trimesters ($\rho = 0.22$, S) and third ($\rho = 0.21$, S).
Ramón et al. 2007 [52]	253 infants born to Spanish women, 2005-2006	Mean 16.5 oz/wk in 2nd and 3rd trimesters.	Cord T-Hg associated with large oily fish, lean fish, and mixed fried fish (all S), but not tuna, sardines, or smoked fish.

Abbreviations: T-Hg, total mercury; MeHg, methyl mercury; S, statistically significant; NS, non-significant; ρ, correlation coefficient; β, beta coefficient. 'Blood' and 'hair' refer to maternal blood and hair.

a. Fish consumption frequencies originally reported in in months or years are reported here as units/wk.

b. Fish consumption originally reported in g/day or g/wk is reported as oz./week here.

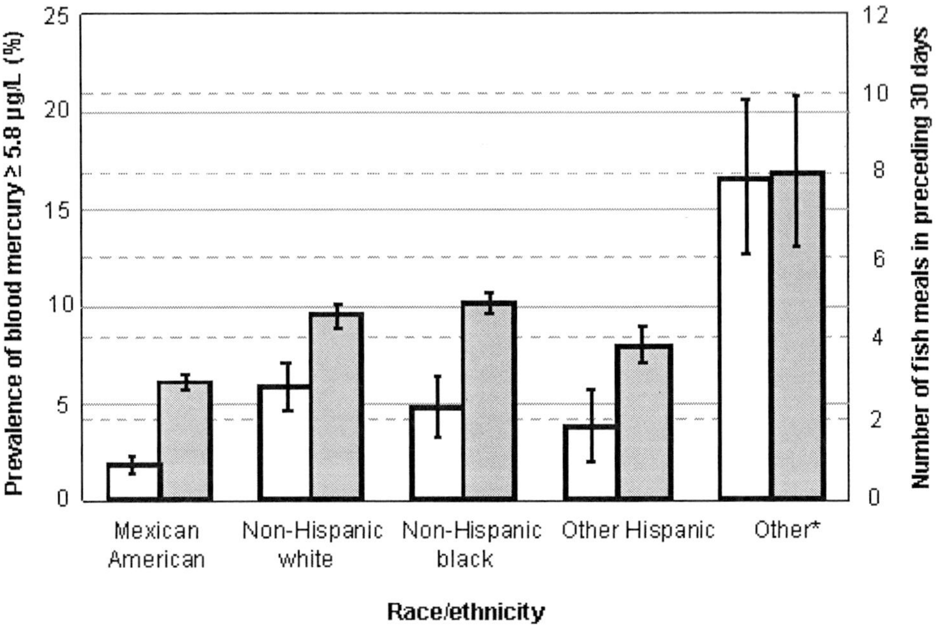

Figure 1. Prevalence of elevated blood mercury and number of fish meals among NHANES female participants aged 16 to 45 years, 1999-2002 (using data from Hightower et al. 2006).*

Prevalence of elevated blood mercury plotted as white columns on left axis; recent fish meal consumption plotted as grey columns on right axis. Bars represent 95% confidence intervals. Other race/ethnicity comprises women who self-identified as Asian, Pacific Islander, Native American, or multiracial.

Hightower's results corroborated those from Mahaffey's previous analysis of NHANES 1990-2000 data [38]. This analysis of 1,709 female participants aged 16-49 years also showed that fish consumption measured by 24-hour recall and blood mercury concentration were positively correlated with age. Fish consumption was highest among women aged 30-39 (geometric mean, 2.0 g/d, 95% CI, 1.5-2.5 g/d) and lowest among women aged 16-19 (geometric mean, 1.5 g/d, 95% CI, 1.2-1.8 g/d). Mean consumption was 1.8 and 1.7 g/d among women aged 20-29 and 40-49, respectively. The prevalence of blood reference dose exceedance increased from 2.4% among 16-19 year olds to 8.8% among 40-49 year olds.

Silver et al. assessed consumption correlates in 458 fish-consuming participants of the Special Supplemental Nutrition Program for Women, Infants, and Children (WIC) in Stockton, California [54]. Vietnamese, other Asian or Pacific Islander, and African American participants had the highest intakes (range of geometric means 10.2–17.6 oz/wk), and white and Native American participants the lowest (range of geometric means 5.4-5.6 oz/wk, overall p-value across racial groups, 0.30). Pregnant women had lower average consumption than non-pregnant women (4.1 vs. 7.4 oz/wk, p=0.0001). As reported by Hightower and Mahaffey, age was directly associated with fish consumption: geometric mean consumption was 6.6 oz/wk in women <17 years of age and 9.9 oz/wk in women ≥50 years (trend p=0.03). Educational attainment and breastfeeding status were not associated with consumption.

Sontrop et al. assessed fish consumption in 2,394 pregnant Ontario women at 10-22 weeks gestation using a validated food frequency questionnaire (FFQ) [55]. Infrequent fish consumption (<1 serving/week) was associated with less than high school education (adjusted risk ratio [RR] 1.60; 95% CI 1.30-1.96), age <21 years (adjusted RR versus ≥35 years, 1.53; 95% CI 1.12-2.02), current smoking (adjusted RR versus non-smoker, 1.20; 1.00-1.43); physical inactivity (adjusted RR versus regularly active, 1.22; 1.06-1.39); and obesity (adjusted RR versus normal weight, 1.19; 1.01-1.41).

Advisory awareness and compliance. Table 2 summarizes studies of the extent and behavioral impacts of advisory awareness in the United States. In these diverse populations, some of which included men, the prevalence of state or federal advisory awareness ranged from 20% to 51% [54, 56-61]. Three studies examined pregnant populations, and just one directly assessed behavioral consequences of awareness. Oken et al. examined fish consumption in Massachusetts before and after the 2001 FDA advisory using FFQs in each trimester [61]. No changes during pregnancy were detected. However, consumption of dark meat fish, canned tuna, and white meat fish (but not shellfish) decreased between 1999 and 2002. Overall consumption declined by about 1.4 servings/month (95% CI, 0.7-2.0) from before to after the advisory. Gliori et al. surveyed 736 new mothers to measure the impact of a Wisconsin fish consumption advisory program targeted to pregnant women [59]. Fewer than 20% participants had seen the program outreach materials. However, 65% were generally aware of mercury's health effects. These participants were asked how they changed their diet after learning about mercury. Forty two percent continued to eat the same amount of fish, 23% ate less, 17% ate different types, <1% ate more, and the remaining 17% continued to abstain from fish. The authors asked where women had seen the outreach materials. The obstetrician's office was the most commonly cited location (33%). Silver et al. administered multilingual interviews to 500 participants of the Special Supplemental Nutrition Program for Women, Infants, and Children (WIC), 66 of whom were pregnant at the time [54]. Forty five percent knew that advisories existed ('general awareness'), while 31% were able to identify specific behaviors consistent with the advisories ('specific awareness'). Women with specific advisory awareness reported lower average fish consumption that those without specific awareness (5.8 versus 7.5 oz/week, p=0.02). However, general awareness was not associated with consumption (6.7 versus 7.0 oz/week, p=0.60). The most common sources of advisory awareness were television (15%), friends or family (9%), health care providers (6%), WIC (6%), and printed media (6%).

An Estimation of Advisory Compliance

Analytical population. We measured periconceptional fish consumption, estimated advisory compliance, and identified non-compliance risk factors among participants of the Omega Study. This is an ongoing prospective cohort primarily designed to examine preeclampsia risk factors. Study procedures have been described elsewhere [62].

Table 2. Published studies of the extent of advisory awareness and its impact on fish consumption behaviors in US populations

Reference	Population	Assessment	Advisory awareness and behavioral impact	Comments
Levine et al. 1999 [60]	396 licensed MI anglers; 22% female, mean age 34	Telephone survey	35% had read advisory. Unclear impact: those who read info were more likely to eat contaminated species but also more likely to engage in exposure-reducing practices (e.g. changing fishing location).	Advisory information was provided in writing at time of licensing.
Beehler et al. 2003 [57]	33 Latino anglers and their partners (mean age 32) in western NY, 1999	Focus groups	51% were aware of local advisories. Those with awareness of poor water quality were more likely to catch and release.	Awareness according to age and gender was not specified.
Oken et al. 2003 [61]	2235 women attending obstetric care in western MA, 1999-2002	Interrupted time-series comparing pre- and post-2001 FDA advisory	Over the study period, consumption of dark meat fish, canned tuna, and white meat fish decreased. There was no change in shellfish intake.	Consumption was measured with 3 FFQs and was averaged over pregnancy.
Anderson et al. 2004 [56]	3,013 women aged 18-45 in 12 states, 1998-1999	Telephone survey	20% were aware of state advisories. 71% were aware of general toxic effects. Behavioral impact of awareness was not assessed.	<50% were aware of which fish characteristics were related to exposure.

Table 2. (Continued)

Reference	Population	Assessment	Advisory awareness and behavioral impact	Comments
Burger 2005 [58]	180 college students and employees in central NJ (mean age 34, 58% women), year not specified	Interview questionnaire	31% were aware of FDA advisories in general, 13% were aware of swordfish/shark component. $\geq$70% were aware of two NJ warnings. Behavioral impact of awareness was not assessed.	There were no gender differences in awareness; 21-45 year olds were less aware than older participants.
Gliori et al. 2006 [59]	736 WI residents delivering 6/2003 (shortly after fish consumption outreach program)	Questionnaire mailed after delivery	<20% had seen outreach information, but 65% were aware of health risks associated with mercury. Of those 65%, 58% had changed fish consumption behaviors after learning of risks.	Obstetrician's office was most cited location at which women saw outreach materials.
Silver et al. 2007 [54]	500 WIC participants in Stockton, CA; pregnant or with young children	Multilingual interview questionnaire at the clinic	45% were aware of advisories and 31% knew of specific recommendations given in advisories. Women with specific awareness consumed less fish than those without (GM 5.8 vs. 7.5 oz/wk).	29% exceeded local or federal advisory limits.

Abbreviations: CA, California; FDA, U.S. Food and Drug Administration; FFQ, food frequency questionnaire; GM, geometric mean; MA, Massachusetts; MI, Michigan; NJ, New Jersey; NY, New York; WI, Wisconsin; WIC, Special Supplemental Nutrition Program for Women, Infants, and Children.

Briefly, English-speaking women aged 18 or older were recruited before 20 weeks gestation from prenatal clinics affiliated with Swedish Medical Center and Tacoma General Hospital in Seattle and Tacoma, Washington. Study procedures were approved by institutional review boards at both institutions. After providing written informed consent, participants were interviewed and given a semi-quantitative FFQ (completed at 14.2 weeks gestation, on average) [63]. Study personnel asked women to report usual consumption in the periconceptional period, i.e. the period spanning the three months before pregnancy and early pregnancy.

The FFQ included items designed to assess usual periconceptional consumption of 35 types of fish and shellfish. This fish questionnaire was designed to assess dietary consumption of PUFAs. Women were prompted to record how often they ate each type by marking one of nine categories, ranging from 'never or less than once per month' to'2+ per day'. Women were prompted to record the usual amount consumed, choosing among small, medium, and large sizes. A reference size was provided for each type. The reference size was 3 oz. for all types except anchovies (1 oz.), fish sticks (4 sticks or 4 oz.), and canned albacore or regular tuna (2.7 oz.).

We examined fish consumption among participants who enrolled between 1996 and 2005. During this period 2,471 women (82% of those enrolled) participated in interviews, completed the fish questionnaire, and were followed through delivery. We excluded 88 participants with missing information regarding maternal characteristics (listed in Table 4). Thus, 2,383 women comprised the analytical population.

Data analysis. We converted participants' fish questionnaire responses to consumption frequencies expressed as 6 oz. servings/week. Our classification of mercury content was based on that used by the FDA, which distinguishes groups of fish 'with highest levels of mercury', those 'with lower levels of mercury', and 'other' groups [64]. We classified 26 types as low-mercury. We classified two species, shark and swordfish, as high-mercury types. Although the FDA also classifies king mackerel and tilefish as having high mercury, we did not assess consumption of these two species individually. However, we assessed consumption of snapper (a term encompassing several species including tilefish) and mackerel (a term encompassing king mackerel and other species). Because most of the species within these groups fall into FDA's 'other' category, we classified them as medium-mercury types. This classification may have caused us to underestimate high-mercury fish consumption. In total, we classified seven fish types as medium-mercury.

We estimated federal advisory compliance using these consumption data. We defined compliance as usual consumption of up to two 6 oz. servings/week, with no more than one serving comprised of medium-mercury fish. We defined non-compliance as excess consumption of low- or medium-mercury fish or any consumption of high-mercury fish. We defined non-consumption as having listed 'never or less than once a month' for all fish types. We were unable to assess local fish consumption.

We used multivariable logistic regression models to estimate adjusted odds ratios (OR) and 95% confidence intervals (CI) for non-compliance and no fish consumption according to maternal characteristics, adjusting for all characteristics simultaneously. We used women classified as compliant as the referent group.

Results. More than 90% of participants consumed fish before and during pregnancy (Table 3). The most common species included fresh salmon, shrimp, halibut, and albacore and light tuna. 3.4% consumed high mercury species (swordfish or shark). Median servings per week among those who consumed fish ranged from 0.85 overall to 0.12 for high-mercury species. Median consumption was significantly and directly associated with the following maternal characteristics: advanced age, Asian race, lower education, higher income, smoking before pregnancy, and regular pre-pregnancy physical activity (Table 4).

As shown in Table 4, 208 women (8.7%) reported no periconceptional fish consumption. Compared with women who complied with advisories, non-consumers were more likely to be 18-20 years old (adjusted OR versus 20-34 years, 2.1; 95% CI, 1.0-4.5). Non-consumption was significantly associated with leanness (adjusted OR versus ideal BMI 1.6; 95% CI, 1.1-2.4) and overweight (adjusted OR 1.6; 95% CI 1.0-2.4) but not obesity (adjusted OR 1.4; 95% CI 0.6-2.1).

Table 3. Periconceptional fish consumption among western Washington cohort study participants: overall and according to fish type and mercury content (N=2,383)

	% of women who ate the fish	Consumption percentiles among women who ate the fish (6 oz. servings/week)				
		25th	50th	75th	95th	Maximum
All types combined	91.3	0.43	0.85	1.43	2.85	27.29
Five most commonly consumed types:						
Fresh salmon	27.29	0.12	0.17	0.29	0.75	4.13
Shrimp	58.2	0.12	0.14	0.29	0.50	4.13
Albacore tuna, canned	39.0	0.10	0.26	0.26	0.90	3.71
Regular tuna, canned	35.8	0.10	0.26	0.26	0.68	3.71
Halibut	43.7	0.12	0.12	0.17	0.43	1.00
High-mercury fish or groups containing high-mercury species:						
Snapper or rockfish[a]	7.7	0.12	0.12	0.17	0.43	4.13
Swordfish	3.2	0.12	0.12	0.12	0.29	1.38
Mackerel[b]	2.0	0.06	0.12	0.17	0.43	2.63
Shark	0.5	0.12	0.12	0.12	0.17	0.17
Fish grouped by mercury content:						
Low mercury	88.2	0.33	0.65	1.09	2.23	18.94
Medium mercury	67.0	0.12	0.26	0.45	0.92	12.34
High mercury[c]	3.4	0.12	0.12	0.17	0.29	1.38

a. A group that includes one high-mercury species, tilefish (*aka* golden snapper).
b. A group that includes one high-mercury species, king mackerel.
c. This group does not include snapper, rockfish, or mackerel: see text.

348 women (14.6%) reported consumption in excess of advisory recommendations (Table 4). Excess consumption was associated with older age (adjusted OR for 35-39 versus 20-34 years, 1.7, 95% CI 1.3-2.2; OR for ≥40 years, 2.0; 95% CI 1.3-3.2) and smoking before pregnancy (adjusted OR versus nonsmokers, 1.3; 95% CI 1.0-1.7). It was moderately, but

imprecisely, associated with Asian race (adjusted OR versus white, 1.3; 95% CI 0.9-2.1). Excess consumption was less frequent among women with ≤12 years education (adjusted OR versus >12 years, 0.4; 95% CI 0.2-1.0). Parous women were less likely to report excess consumption than nulliparous women (adjusted OR, 0.6; 95% CI, 0.5-0.8).

Table 4. Median fish consumption (6 oz. servings/week) and adjusted odds ratios and 95% confidence intervals for fish consumption behaviors according to maternal characteristics (n=2,383)[a]

		Consumption behaviors				
		Compliance[b]	No consumption		Non-compliance	
	Median	(n=1,827)	(n=208)		(n=348)	
Characteristic	consumption	n	n	Adj. OR (CI)	n	Adj. OR (CI)
Age (years)						
18-20**	0.35	53	17	2.1 (1.0-4.5)	6	0.8 (0.2-2.2)
21-34[c]	0.73	1229	141	1.0	197	1.0
35-39**	0.87	449	41	0.8 (0.5-1.1)	114	1.7 (1.3-2.2)
≥40**	0.88	96	9	0.8 (0.4-1.5)	31	2.0 (1.3-3.2)
Race/ethnicity						
White[c]	0.74	1576	181	1.0	298	1.0
Black	0.66	25	4	0.7 (0.2-2.3)	5	1.3 (0.5-3.5)
Hispanic	0.68	54	9	1.1 (0.5-2.4)	10	1.2 (0.6-2.4)
Asian*	0.94	119	13	0.9 (0.5-1.6)	29	1.3 (0.9-2.1)
Native/other	0.77	53	1	0.1 (0.02-1.0)	7	0.8 (0.3-1.7)
Parity						
Nulliparous[c]	0.78	1145	138	1.0	246	1.0
Parous	0.70	682	70	0.9 (0.6-1.2)	102	0.6 (0.5-0.8)
Marital status						
Married[c]	0.76	1688	181	1.0	322	1.0
Unmarried	0.66	139	27	1.3 (0.7-2.3)	26	1.0 (0.6-1.5)
Employment in early pregnancy						
Employed[c]	0.74	1487	171	1.0	282	1.0
Unemployed	0.77	340	37	0.9 (0.6-1.4)	66	1.2 (0.9-1.7)
Education (years)						
≤12**	0.33	70	16	1.6 (0.8-3.0)	5	0.4 (0.2-1.0)
>12[c]	0.77	1757	192	1.0	343	1.0
Household income (US$)						
<30,000**	0.46	34	8	1.0 (0.4-2.7)	4	0.7 (0.2-2.2)
30-69,000**	0.68	413	47	0.9 (0.6-1.3)	72	0.9 (0.7-1.3)
≥70,000[c]	0.80	1355	146	1.0	268	1.0
Smoking history						
Never[c]	0.73	1340	144	1.0	241	1.0
Before preg*	0.85	386	46	1.1 (0.8-1.6)	89	1.3 (1.0-1.7)
During preg	0.63	101	18	1.1 (0.6-2.1)	18	1.2 (0.7-2.1)

Table 4. (Continued)

		Consumption behaviors				
		Compliance[b]	No consumption		Non-compliance	
	Median	(n=1,827)	(n=208)		(n=348)	
Characteristic	consumption	n	n	Adj. OR (CI)	n	Adj. OR (CI)
Pre-pregnancy body mass index (kg/m^2)						
≤20	0.74	320	48	1.6 (1.1-2.4)	71	1.2 (0.9-1.7)
20-24[c]	0.76	1073	99	1.0	193	1.0
25-29	0.73	286	41	1.6 (1.1-2.4)	59	1.1 (0.8-1.6)
≥30	0.81	148	20	1.4 (0.6-2.1)	25	1.0 (0.7-1.7)
Regular physical activity in the year before pregnancy						
Yes[c]	0.77	1678	191	1.0	329	1.0
No*	0.56	149	20	0.9 (0.5-1.6)	19	0.7 (0.4-1.2)

a. All characteristics are included as covariates in each model.

b. The referent (compliance) group consists of women whose self-reported periconceptional fish consumption met FDA/EPA guidelines.

c. Referent maternal characteristic category.

* p for Wilcoxon rank-sum test <0.05.

** p for Wilcoxon rank-sum test <0.001.

Conclusion

A fairly large body of literature shows that maternal fish consumption is related to mercury concentrations in both maternal and infant tissues. Future studies should measure and describe these relations in more detail. Such results would provide information that may prove useful in targeting and modifying advisories. For instance, with two exceptions, studies have not provided estimates of the magnitude of mercury exposure associated with each serving increase in consumption [3, 38]. Furthermore, there is little information regarding the species that are most strongly related to maternal and fetal body burdens. While larger predatory species contain the highest concentrations of methylmercury, more commonly consumed species such as albacore tuna may contribute more to exposures at the population level. If we identify the species that most commonly influence maternal and fetal mercury exposures, we may be able to target advisories to women who regularly consume them.

Here we report results from the first analysis, to our knowledge, to assess risk factors for a lack of fish consumption and consumption in excess of the FDA/EPA advisory. Analyses of NHANES survey data have provided generalizable estimates of the relations between race/ethnicity and age and fish consumption [38, 53]. Like previous investigators, we found that Asian women were more likely to report high consumption, though the precision of this result was hampered by small numbers [38, 53, 54]. Two other localized studies of Californians and Ontarians also show direct associations between age and fish consumption, which our findings corroborate [54, 55]. Like Sontrop et al., we found that low education and young age were associated with low consumption, though we did not observe an association with physical inactivity [55]. Our data from western Washington residents also suggest that BMI and nulliparity are related to fish consumption behaviors. We encourage further analyses to identify risk factors of excess fish consumption in diverse populations. Just as importantly, such analyses should also identify women who consume little or no fish before and during pregnancy. Results from this research could inform interventions designed to encourage both groups to eat fish in moderate quantities.

We observed a strong association between nulliparity and risk of excess periconceptional fish consumption that was independent of age and other maternal characteristics. Though we could not measure advisory awareness directly, this association suggests to us that women may first learn of advisories or the risks of excess consumption during prenatal care. Our hypothesis is supported by the survey conducted by Gliori et al., in which women most commonly reported the obstetrician's office as the source of advisory materials [59]. Further examination of the timing of advisory awareness and subsequent behavioral change (if any) is critical. This timing likely influences the extent to which behavioral changes may reduce the fetal risks of mercury exposure. Methylmercury's elimination half-life in human blood is 70 to 80 days [15]. Given this persistence, reduction of fish consumption during early pregnancy may not meaningfully impact fetal exposures during potentially critical developmental periods in early- and mid-gestation. The influences of behavioral change timing may be clarified by carefully measuring maternal awareness and fish consumption behaviors and directly relating them to biological mercury concentrations. Collection of maternal hair and blood samples during early and mid-pregnancy, and infant cord blood after delivery, could also maximize investigators' ability to assess distinct time intervals of mercury exposure.

Limited data suggest that only half of American reproductive-aged women, at most, are aware of the fish consumption advisories targeted to them [54, 56-61]. The scant evidence also suggests that advisory awareness does not have a strong impact on fish consumption [54, 59]. Until we better understand the impacts of federal and state advisories, we will be unable to improve the ways in which we communicate the benefits and risks of fish consumption to reproductive-aged women. Surveys of knowledge, attitudes, and practices related to fish consumption are needed, as well as studies designed to assess characteristics associated with advisory knowledge and behavioral change. Future studies should also measure the extent and timing of behavioral change in relation to the timing of pregnancy to ensure that maternal behavioral changes benefit infants.

References

[1] Cohen JT, Bellinger DC, Connor WE, Shaywitz BA. A quantitative analysis of prenatal intake of n-3 polyunsaturated fatty acids and cognitive development. *Am. J. Prev .Med.* 2005 Nov; 29, 366-74.

[2] Institute of Medicine (U.S.). Committee on Nutrient Relationships in Seafood: Selections to Balance Benefits and Risk, Nesheim MC, Yaktine AL, National Academies Press (U.S.). Seafood choices: balancing benefits and risks. 2007 [cited; xiv, 722 p.]. Available from: http://www.nap.edu/books/0309102189/html/index.html

[3] Oken E, Wright RO, Kleinman KP, Bellinger D, Amarasiriwardena CJ, Hu H, et al. Maternal fish consumption, hair mercury, and infant cognition in a U.S. Cohort. *Environ. Health Perspect.* 2005 Oct; 113, 1376-80.

[4] Cetin I, Koletzko B. Long-chain omega-3 fatty acid supply in pregnancy and lactation. *Curr. Opin. Clin. Nutr. Metab. Care.* 2008 May; 11, 297-302.

[5] Oken E, Radesky JS, Wright RO, Bellinger DC, Amarasiriwardena CJ, Kleinman KP, et al. Maternal fish intake during pregnancy, blood mercury levels, and child cognition at age 3 years in a US cohort. *Am. J. Epidemiol.* 2008 May 15; 167, 1171-81.

[6] Cohen JT, Bellinger DC, Shaywitz BA. A quantitative analysis of prenatal methyl mercury exposure and cognitive development. *Am. J. Prev. Med.* 2005 Nov; 29, 353-65.

[7] Grandjean P, White RF. Effects of methylmercury exposure on neurodevelopment. *Jama.* 1999 Mar 10; 281, 896; author reply 7.

[8] Sorensen N, Murata K, Budtz-Jorgensen E, Weihe P, Grandjean P. Prenatal methylmercury exposure as a cardiovascular risk factor at seven years of age. *Epidemiology.* 1999 Jul; 10, 370-5.

[9] Oily fish intake during pregnancy - association with lower hyperactivity but not with higher full-scale IQ in offspring. *J Child Psychol Psychiatry.* 2008 Apr 15.

[10] Centers for Disease Control and Prevention. Blood mercury levels in young children and childbearing-aged women--United States, 1999-2002. *MMWR Morb Mortal Wkly Rep.* 2004 Nov 5; 53, 1018-20.

[11] United States. Environmental Protection Agency. Office of Water. National listing of fish advisories: fact sheet. 2004 [cited; 6 p.]. Available from: http://purl.access.gpogov/GPO/LPS67594

[12] Levenson CW, Axelrad DM. Too much of a good thing? Update on fish consumption and mercury exposure. *Nutr Rev.* 2006 Mar; 64, 139-45.

[13] Marien K, Stern AH. An examination of the trade-offs in public health resulting from the use of default exposure assumptions in fish consumption advisories. *Environ. Res.* 2005 Jun; 98, 258-67.

[14] Smith KM, Sahyoun NR. Fish consumption: recommendations versus advisories, can they be reconciled? *Nutr Rev.* 2005 Feb; 63, 39-46.

[15] National Research Council (U.S.). Committee on the Toxicological Effects of Methylmercury. Toxicological effects of methylmercury. Washington, DC: National Academy Press 2000.

[16] United States. Environmental Protection Agency. Office of Air Quality Planning and Standards., United States. Environmental Protection Agency. Office of Research and Development. Mercury study report to Congress [microform .]. [Research Triangle Park, N.C.]: Office of Air Quality Planning and Standards and Office of Research and Development, U.S. Environmental Protection Agency,; 1997.

[17] Cernichiari E, Myers GJ, Ballatori N, Zareba G, Vyas J, Clarkson T. The biological monitoring of prenatal exposure to methylmercury. *Neurotoxicology.* 2007 Sep; 28, 1015-22.

[18] United States. Environmental Protection Agency. Office of Water. Backgrounder for the 2004 FDA/EPA consumer advisory what you need to know about mercury in fish

and shellfish. 2004 [cited; 5 p.]. Available from: http://purl.access.gpo.gov/GPO/LPS67598
[19] U.S. Department of Health and Human Services and U.S. Environmental Protection Agency. Mercury levels in commercial fish and shellfish. May 2001 February 2006 [cited 10/03/07]; Available from: http://www.cfsan.fda.gov/~frf/sea-mehg.html
[20] Burger J, Gochfeld M. Mercury in canned tuna: white versus light and temporal variation. *Environ Res.* 2004 Nov; 96, 239-49.
[21] Spurgeon A. Prenatal methylmercury exposure and developmental outcomes: review of the evidence and discussion of future directions. *Environ Health Perspect.* 2006 Feb; 114, 307-12.
[22] Grandjean P, Weihe P, White RF, Debes F, Araki S, Yokoyama K, et al. Cognitive deficit in 7-year-old children with prenatal exposure to methylmercury. *Neurotoxicol. Teratol.* 1997 Nov-Dec; 19, 417-28.
[23] Kjellstrom T, Kennedy P, Wallis S, Stewart A, Friberg L, Lind B. Physical and Mental Development of Children with Prenatal Exposure to Mercury from Fish. Solna, Sweden: National Swedish Environmental Protection Board; 1989.
[24] Myers GJ, Davidson PW, Cox C, Shamlaye CF, Palumbo D, Cernichiari E, et al. Prenatal methylmercury exposure from ocean fish consumption in the Seychelles child development study. *Lancet.* 2003 May 17; 361, 1686-92.
[25] Jedrychowski W, Perera F, Jankowski J, Rauh V, Flak E, Caldwell KL, et al. Fish consumption in pregnancy, cord blood mercury level and cognitive and psychomotor development of infants followed over the first three years of life: Krakow epidemiologic study. *Environ. Int.* 2007 Nov; 33, 1057-62.
[26] Davidson PW, Myers GJ, Weiss B, Shamlaye CF, Cox C. Prenatal methyl mercury exposure from fish consumption and child development: a review of evidence and perspectives from the Seychelles Child Development Study. *Neurotoxicology.* 2006 Dec; 27, 1106-9.
[27] Grandjean P, Budtz-Jorgensen E, Jorgensen PJ, Weihe P. Umbilical cord mercury concentration as biomarker of prenatal exposure to methylmercury. *Environ. Health Perspect.* 2005 Jul; 113, 905-8.
[28] Grandjean P, Weihe P. Neurobehavioral effects of intrauterine mercury exposure: potential sources of bias. *Environ. Res.* 1993 Apr; 61, 176-83.
[29] Hubbs-Tait L, Nation J, Krebs N, Bellinger D. Neurotoxicants, micronutrients, and social environments: individual and combined effects on children's development. *Psychol. Sci. Publ. Interest.* 2005 2005; 6, 57-121.
[30] Olsen SF, Osterdal ML, Salvig JD, Weber T, Tabor A, Secher NJ. Duration of pregnancy in relation to fish oil supplementation and habitual fish intake: a randomised clinical trial with fish oil. *Eur. J. Clin. Nutr.* 2007 Aug; 61, 976-85.
[31] Similar effects on infants of n-3 and n-6 fatty acids supplementation to pregnant and lactating women. *Pediatrics.* 2001 Nov; 108, E82.
[32] Muthayya S, Dwarkanath P, Thomas T, Ramprakash S, Mehra R, Mhaskar A, et al. The effect of fish and omega-3 LCPUFA intake on low birth weight in Indian pregnant women. *Eur. J. Clin. Nutr.* 2007 Oct 24.

[33] Halldorsson TI, Meltzer HM, Thorsdottir I, Knudsen V, Olsen SF. Is high consumption of fatty fish during pregnancy a risk factor for fetal growth retardation? A study of 44,824 Danish pregnant women. *Am J Epidemiol.* 2007 Sep 15; 166, 687-96.

[34] American Heart Association. Fish and omega-3 fatty acids: AHA recommendation. 2004 [cited July 7, 2008]; Available from: http://www.americanheart.org /presenter.jhtml?identifier=4632

[35] Bjornberg KA, Vahter M, Petersson-Grawe K, Glynn A, Cnattingius S, Darnerud PO, et al. Methyl mercury and inorganic mercury in Swedish pregnant women and in cord blood: influence of fish consumption. *Environ. Health Perspect.* 2003 Apr; 111, 637-41.

[36] United States. Environmental Protection Agency. Office of Water. National listing of fish advisories: general fact sheet: 2005/06 national listing. 2007 [cited; 6 p.]. Available from: http://www.epa.gov/waterscience/fish/advisories/2006/index.html

[37] Stern AH, Korn LR, Ruppel BE. Estimation of fish consumption and methylmercury intake in the New Jersey population. *J. Expo. Anal. Environ .Epidemiol.* 1996 Oct-Dec; 6, 503-25.

[38] Mahaffey KR, Clickner RP, Bodurow CC. Blood organic mercury and dietary mercury intake: National Health and Nutrition Examination Survey, 1999 and 2000. *Environ Health Perspect.* 2004 Apr; 112, 562-70.

[39] Trasande L, Landrigan PJ, Schechter C. Public health and economic consequences of methyl mercury toxicity to the developing brain. *Environ Health Perspect.* 2005 May; 113, 590-6.

[40] Arnold SM, Lynn TV, Verbrugge LA, Middaugh JP. Human biomonitoring to optimize fish consumption advice: reducing uncertainty when evaluating benefits and risks. *Am J Public Health.* 2005 Mar; 95, 393-7.

[41] Crump KS, Van Landingham C, Shamlaye C, Cox C, Davidson PW, Myers GJ, et al. Benchmark concentrations for methylmercury obtained from the Seychelles Child Development Study. *Environ. Health Perspect.* 2000 Mar; 108, 257-63.

[42] Groth E, 3rd. RE: "MATERNAL FISH INTAKE DURING PREGNANCY, BLOOD MERCURY LEVELS, AND CHILD COGNITION AT AGE 3 YEARS IN A US COHORT". *Am. J. Epidemiol.* 2008 Jun 12.

[43] National Library of Medicine. PubMed [online database]. 2008.

[44] Kim EH, Kim IK, Kwon JY, Kim SW, Park YW. The effect of fish consumption on blood mercury levels of pregnant women. *Yonsei. Med. J.* 2006 Oct 31; 47, 626-33.

[45] Morrissette J, Takser L, St-Amour G, Smargiassi A, Lafond J, Mergler D. Temporal variation of blood and hair mercury levels in pregnancy in relation to fish consumption history in a population living along the St. Lawrence River. *Environ. Res.* 2004 Jul; 95, 363-74.

[46] Muckle G, Ayotte P, Dewailly E, Jacobson SW, Jacobson JL. Determinants of polychlorinated biphenyls and methylmercury exposure in inuit women of childbearing age. *Environ. Health Perspect.* 2001 Sep; 109, 957-63.

[47] Sato RL, Li GG, Shaha S. Antepartum seafood consumption and mercury levels in newborn cord blood. *Am. J. Obstet. Gynecol.* 2006 Jun; 194, 1683-8.

[48] Stern AH, Gochfeld M, Weisel C, Burger J. Mercury and methylmercury exposure in the New Jersey pregnant population. *Arch. Environ. Health.* 2001 Jan-Feb; 56, 4-10.

[49] Vahter M, Akesson A, Lind B, Bjors U, Schutz A, Berglund M. Longitudinal study of methylmercury and inorganic mercury in blood and urine of pregnant and lactating women, as well as in umbilical cord blood. *Environ. Res.* 2000 Oct; 84, 186-94.

[50] Cordier S, Grasmick C, Paquier-Passelaigue M, Mandereau L, Weber JP, Jouan M. Mercury exposure in French Guiana: levels and determinants. *Arch. Environ. Health.* 1998 Jul-Aug; 53, 299-303.

[51] Gibicar D, Horvat M, Nakou S, Sarafidou J, Yager J. Pilot study of intrauterine exposure to methylmercury in Eastern Aegean islands, Greece. *Sci. Total Environ.* 2006 Aug 31; 367, 586-95.

[52] Ramon R, Murcia M, Ballester F, Rebagliato M, Lacasana M, Vioque J, et al. Prenatal exposure to mercury in a prospective mother-infant cohort study in a Mediterranean area, Valencia, Spain. *Sci. Total Environ.* 2008 Mar 15; 392, 69-78.

[53] Hightower JM, O'Hare A, Hernandez GT. Blood mercury reporting in NHANES: identifying Asian, Pacific Islander, Native American, and multiracial groups. *Environ. Health Perspect.* 2006 Feb; 114, 173-5.

[54] Silver E, Kaslow J, Lee D, Lee S, Lynn Tan M, Weis E, et al. Fish consumption and advisory awareness among low-income women in California's Sacramento-San Joaquin Delta. *Environ. Res.* 2007 Jul; 104, 410-9.

[55] Sontrop JM, Campbell MK, Evers SE, Speechley KN, Avison WR. Fish consumption among pregnant women in London, Ontario: associations with socio-demographic and health and lifestyle factors. *Can. J. Public Health.* 2007 Sep-Oct; 98, 389-94.

[56] Anderson HA, Hanrahan LP, Smith A, Draheim L, Kanarek M, Olsen J. The role of sport-fish consumption advisories in mercury risk communication: a 1998-1999 12-state survey of women age 18-45. *Environ. Res.* 2004 Jul; 95, 315-24.

[57] Beehler GP, McGuinness BM, Vena JE. Characterizing Latino anglers' environmental risk perceptions, sport fish consumption, and advisory awareness. *Med. Anthropol. Q.* 2003 Mar; 17, 99-116.

[58] Burger J. Fishing, fish consumption, and knowledge about advisories in college students and others in central New Jersey. *Environ. Res.* 2005 Jun; 98, 268-75.

[59] Gliori G, Imm P, Anderson HA, Knobeloch L. Fish consumption and advisory awareness among expectant women. *Wmj.* 2006 Mar; 105, 41-4.

[60] Levine KJ, Miller VD, Kamrin MA, Dearing JW. Anglers' attitudes, beliefs, and behaviors as impacted by the Michigan Fish Consumption Advisory. *J. Public Health. Manag Pract.* 1999 Nov; 5, 18-28.

[61] Oken E, Kleinman KP, Berland WE, Simon SR, Rich-Edwards JW, Gillman MW. Decline in fish consumption among pregnant women after a national mercury advisory. *Obstet. Gynecol.* 2003 Aug; 102, 346-51.

[62] Butler CL, Williams MA, Sorensen TK, Frederick IO, Leisenring WM. Relation between maternal recreational physical activity and plasma lipids in early pregnancy. *Am. J. Epidemiol.* 2004 Aug 15; 160, 350-9.

[63] Patterson RE, Kristal AR, Tinker LF, Carter RA, Bolton MP, Agurs-Collins T. Measurement characteristics of the Women's Health Initiative food frequency questionnaire. *Ann. Epidemiol.* 1999 Apr; 9, 178-87.

[64] U.S. Department of Health and Human Services and U.S. Environmental Protection Agency. Mercury levels in commercial fish and shellfish. 2001 February 2006 [cited 10/03/07]; Available from: http://www.cfsan.fda.gov/~frf/sea-mehg.html

In: Fish Consumption and Health
Editors: George P. Gagne and Richard H. Medrano
ISBN 978-1-60741-151-2

Chapter IX

Emerging Health Risk Due to Shellfish Consumption?

Juan Martínez-Hernández[*,1, 2]**, *Victoria Lopez-Rodas***[2] **and *Eduardo Costas***[2]

[1] Servicio de Medicina Preventiva y Salud Pública. Hospital Carlos III 28029, Madrid, Spain

[2]Genética, Facultad de Veterinaria, Universidad Complutense, 28040, Madrid, Spain

Abstract

Shellfish is one of the main commercial productions in aquaculture industry of USA, Europe, Japan and China. Since shellfish feed by filtration, often they accumulate (bio-) toxins, bacteria, and parasites. As a result thousands people suffered Paralytic-, Amnesic-, or Diarrheic Shellfish Poisoning (PSP, ASP, DSP), ciguatera and other diseases during the twentieth century (and as consequence many people died). Relevant efforts have been made in prevention programs, treatments and legislation to reduce the incidence of such diseases (usually obtaining satisfactory results).

However, new emerging health problems are being linked to shellfish consumption. For example, mussels (Mytilus galloprovincialis) and cockles (Cerastoderma edule) store infectious Cryptosporidium parvum oocyst, which produce diarrhea in humans (being a relevant cause of died in AIDS infected people). On the other hand, European legislation allows less than 0,16μg DSP toxin (okadaic acid equivalent) per gram of meat (Decision 2002/225/CE), which is enough to prevent Diarrheic Shellfish Poisoning. In spite of this, DSP toxins are also potent tumor promoters. An epidemiological study shows a statistically significant correlation ($p<0.001$) between consumption of molluscs and the incidence of colorectal cancer (an increase of 7 times in shellfish consumption produced

* Corresponding Address: Dr. Juan Martínez Hernández, Servicio de Medicina Preventiva, Hospital Carlos III, c/ Sinesio Delgado 10, 28029 Madrid (Spain). jmartinez.hciii@salud.madrid.org, +34660632167, +34914532611.

duplication in the risk ratio of colorectal cancer. Many other risks have not been taken in account. For instance, shellfish from estuaries can accumulate microcystins (potent hepatotoxins) produced by freshwater cyanobacteria transported to the estuary by the river. Microcystins have been implicated in the high rates of hepatic cancer in China population.

As a conclusion, a lot of work is still necessary to assure a safety in shellfish consumption.

An important and ongrowing nourishing source is Shellfish. Represents a protein source (11.9% of the edible portion of the mussel is protein) quite a low calorie foodstuff (86 Kcal/100g of the edible portion of the mussel) and is high in Iodine, selenium and Vitamin A[1].

Is a product with a high profitability; probably is the group of foodstuffs from marine origin with more added value in relation with its production costs.

Furthermore, in Spain aquaculture is strongly backed by the authorities (Galicia, in Spain, is the top region in the European rank of shellfish producers, producing regularly 250,000T of mussels, only bested by China with approximately 450,000T per annum) amongst other reasons for these being is that this allows the stability and the traditional ways of life of the communities linked to the sea.

It is a general belief that is recommended to evolve from an extractive society to an intensive fish-farming one. However, intensifying marine fishery has its own risks and disadvantages, especially if we consider that this animals feed by filtrating the plankton present in the seawater and in the estuaries.

Residues from small villages -in which having a water-purification plant for the sewage water is not compulsory- go directly to the seawater or to the estuary. Also main rivers carry in their waters the legal and illegal dumping into the stream of substances. There is also pouring of substances coming from the aquaculture activities, the cleansing of the bilge of vessels and the naval accidents. Sadly, not an uncommon event. Like those of the oil tankers Exxon Valdez (Alaska, 1999) or Prestige (Galicia, 2002).

Chemical substances in the water can be of great importance because they affect to life and reproduction of molluscs and also because these substances can bio-accumulate in the muscle and the hepatopancreas of the animal. This will be the case for the chlorinated organic compounds like the polychlorinated biphenyl (PCB)[2] that is widely used although an illegal substance, among other reasons because of its long lasting toxic effect, agricultural pesticides[3], dioxins[4], benzopyrene[5] and other polycyclic aromatic hydrocarbons from fires, burning of cereal stubble or incineration of residues, that can be accumulated in the animal's fat[6].

Toxic substances affect not only culture species but also wild ones: PCBs can induce an increase of the male population consequently damaging the preservation of the wild species.

Heavy metals (lead, cadmium and mercury for example) are equally absorbed by the mollusks[7] and can accumulate them in the edible part and in the shell, with the subsequent risk if these shells are used to extract the calcium for pharmaceutical use in humans or livestock or to reinforce food. In fact, mussels also accumulate beneficial oligoelements in low doses like selenium.

The risk of neurotoxicity in people eating fish or shellfish is significantly higher if metals like mercury for example, go through an organification process to methylmercury (remember the 1953 outbreak in Minamata Bay, Japan[8, 9]).

Equally, some microorganisms like *Vibrio parahaemolyticus* can poison the meat of bivalves and be the reason for diarrhoea if consumed raw or partially cooked[10]. Cysts of the small parasite *Cryptosporidium parvum* can be retained in shellfish[11]. If consumed by healthy people it can cause diarrhoea and in patients with AIDS chronic diarrhoea.

Biotoxins produced by algae and dinoflagelate must be considered apart due to its severity, its progressive increase in relation with climate change and its possible long-term effect. Following is a brief description of the toxicity of the main chemical species recognized to date, in animal models and in-vitro cultures

It is important to emphasize that the syndromes recognized to date due to shellfish poisoning in relation with verified human health problems are mainly acute. These are PSP (Paralytic Shellfish Poisoning), ASP (Amnesic Shellfish Poisoning), NSP (Neurotoxic Shellfish Poisoning), and DSP (Diarrheic Shellfish Poisonig).

All of them are diagnosed clinically by the knowledge of bivalve consumption in the previous 24 hours. This is because we lack of biomarkers that confirm the exposure to the toxins. We also lack an antitoxin and treatment is symptomatic and life support based.

The most common toxic syndrome of marine biotoxins is ciguatera. It is related to the consumption in tropical regions of barracuda and other big fishes that can accumulate ciguatoxin.

Intoxication with tetrodotoxin (associated with the consumption of pufferfish in Japan[12], and maybe done on purpose in voodoo acts in Haiti[13]) will not be considered in this chapter dedicated to shellfish.

We will neither consider the intoxication with gempylotoxin due to escolar and oilfish consumption, or the scombrotoxin, that forms in fish after captured and kept without chilling, this causes the bacteria to attack the histidin and change it into histamine (histamine poisoning)[14].

Although it has been of interest the acute nature of the these intoxications, evidences are increasingly gathered that show that lipidic and thermoresistant biotoxins, like okadaic acid and its derivatives, could be retained for long periods and even establish an enterohepatic recirculation cycle.

In theory, there is a possibility that DSP toxins can be less toxic if consumed in quantities above the legal limit, in this case they will be virtually totally eliminated by the mild and self-limiting diarrhoeic syndrome itself. If consumed below the limit, the toxin will accumulate in the body fat and circulate now and again, with the subsequent cytotoxic damage, mutagen effect and above all luminal tumour promoter over premalign lesions like the colon polyps. In fact, it is known from old observations that cathartic agents used in the first half of 20th century, compounds made from mercury called calomel, were very poisonous when consumed in low doses and innocuous if taken at the laxative dose, because they were totally removed.

To achieve a correct diagnose a hypothetic biomarker should be sought. Preferably in urine because toxic substances that accumulate in the bile usually are eliminated in the faeces or via kidneys as glucuroconjugates.

As a hypothesis, and with some preliminary epidemiological data available, okadaic acid and its derivatives is a new risk factor for the large bowel cancer[15,16] and maybe for other digestive tract tumors[17].

PSP (Paralytic Shellfish Poisoning)

There are approximately 27 known toxins with the capability to cause PSP. Saxitoxin (STX) is a non-sulphated variation and the most powerful of tem all. In sea water it is produced by dinoflagellates while in fresh water cianobacteria can produce it. Other PSP toxins are the gonyautoxins (GTX).

Signs and symptoms are predominantly neurologic and include tingling, burns, numbness, sleepiness, disjointed speech, and respiratory paralysis that can cause death. All of this commences between half an hour and two hours after consuming the contaminated food and generally disappears after 12 hours unless an acute respiratory failure scenario appears and is not properly treated in hospital or a cardiac arrest due to hypotension.

It is very illustrative of the clinical presentation[18] an episode well described, as the one that took place in Alaska the 6th of June 1990, when 6 fishermen consumed some mussels well cooked. Between one and two hours after eating, 5 of them felt some numbness in the mouth, two of them tingling in the tongue and one numbness in the throat. Four of the fishermen threw up and four had paresthesia in the limbs. All of them were seen in the emergency unit of a hospital. Two of them required several day hospitalizations, but in the end all of them recovered from the episode. The toxin was present in the raw mussels in a higher concentration than in the cooked (during 90 minutes) ones. This is consistent with the fact that is a hydrosoluble and thermoresistant substance that was partially diluted in the stock.

Previously there was a terrible incident in the pacific coast of Guatemala, in Champerico, between July and August 1987. 187 people got sick of PSP and 26 died (14%)[19]. The death rate among children under 6 was 50% and amongst over 18 was 7%, probably due to a less distribution volume of the toxin in children. It was related to the consumption of a wild clam *Amphichaena kindermani* harvested in the beaches. Probably many lives could have been saved with intensive care in these patients.

In the northeast coast of Margarita island, between August and October 1991 there was an important bloom of dinoflagellates *Prorocentrum gracile*, *Gymnodinium catenatum* and *Alexandrium tamarense* that caused 24 cases of PSP from which 16 required hospitalization. The concentration of saxitoxin in mussels (*Perna perna*) exceeded significantly the considered safe level of 80 μg per gram of meat[20].

Episodes of PSP can occur in conjunction with other intoxications[21], as was discovered after analysing the samples obtained from a red tide occurred in Galicia, Spain in 1993, where it was found STX, GTX2, GTX3 and diarrhoeic toxins jointly.

PSP would probably be the biggest acute toxicity problem in this type of foodstuff if shellfish from producing areas were not controlled as they are.

Even in 2003, in a study in Shanghay markets, 8 out of 66 samples had saxitoxin and where positive to mouse bioassay[22].

ASP (Amnesic Shellfish Poisoning)

Several algae species, especially diatomea of the genus *Pseudo-nitzschia* produce a tricarboxilic aminoacid analogous to glutamate called domoic acid capable of serious intoxications.

It seems that his action is exerted by the glutamate receptors like the one for kainite, another excitative aminoacid, especially of the hippocampus area[23]. In the next 24 hours after eating the contaminated molluscs gastrointestinal symptoms of moderate importance start, and the mental symptoms in 48 hours. The neurodegenerative effect is seen principally as loss of the short-term memory, if recovered it takes a long time, resembling in some way to Alzheimer disease, but hallucinations, coma, and death can also occur. Elderly people are especially sensible and all the deaths registered where in this group of age.

In 1987 in Prince Edwards Island (Canada) more than 150 people were intoxicated and 4 died after eating mussels contaminated with domoic acid[24].

Domoic acid was first detected in molluscs in 1998 in Scotland[25] and France[26], and in 1999 in Ireland[27], with a particularly high content in the scallops (*Pecten maximus*) harvested in the east and south coast of the Republic of Ireland (2,820 micrograms per gram of the edible portion). In 2003 and 2004 in Scotland ASP toxins where the most frequent biotoxins in shellfish, found in 69% of the samples in one study[28].

Domoic acid can enter the trofic chain and bioaccumulate in the viscera of *Sepia officinalis*[29], and transfer to marine mammals that feed on them and probably to other cephalopoda, but it does not affect the mantle and because of that toxicity in humans is not expectable, unless young animals from this species are consumed without evisceration, what happens in certain cultures. Deaths in pelicans and sea lions have been registered along side the west coast of the United States in the 90´s. Whales feed on fish and cephalopoda that are contaminated with domoic acid[30], and to date there are no data about toxicity in this animals.

In Europe and Canada[31] the total content of Amnesic Shellfish Poison (ASP) in the edible parts of molluscs (the entire body or any edible part separately) must not exceed 20 micrograms of domoic acid per gram using the HPLC method.

A control system model in Chile, keeps watch on the presence of *Psudonitzchia australis*[32]. The meat of the molluscs is controlled if in the phitoplanckton samples the algae goes above the 65%. If the level of domoic acid rises over 5 micrograms per gram, the extraction area is closed and the molluscs and phitoplanckton is intensively controlled.

NSP (Neurotoxic Shellfish Poisoning)

Gymnodinium breve (*Karenia brevis*) is a marine dinoflagellate that produce brevetoxins, the potent toxins that cause neurotoxic shellfish poisoning[33]. The effects of brevetoxins are not only seen after consumption, but they also have the striking attribute of toxicity by contact with the conjunctival, oropharyngeal and respiratory tract mucosae, causing irritation and allergies. Occasionally they can cause bronchospasm and asmatic crises[34].

In the Genoa outbreak between the 17th and 26th of July 2005, 205 patients were seen in the emergency units (maybe the number was even higher but they did not require specific medic care). An inclusion criterion was to be present in the vicinity of the coast during the bloom. They had fever (64%), sore throat (50%), cough (40%), and in less amount dyspnea, cephalalgy, rhinorrea and vomits. 20% required hospitalization but in the majority of them the symptoms disappeared spontaneously in few hours. Average delay in the onset of symptoms was 7 hours, varying from 30 minutes to 23 hours.

In this event bloom was originated by *Ostreopsis ovata* that produce palytoxin. It is one of the strongest nonpeptidic toxins. Causes cytolysis as a result of the inhibition of Na^+, K^+-ATPase and has been recognised as the causative agent of fatal human seafood poisoning[36].

DSP (Diarrhetic Shellfish Poisonig)

Depending on the amount of okadaic acid (OA) or its derivates, dinophysistoxins (DTXs), pectenotoxins (PTXs), azaspiracids (AZAs), yessotoxins (YTXs), spirolides (SPXs) and gymnodimines (GYMs) ingested, within 30 minutes and 3 hours after, a self-limiting diarrhoea condition starts. Generally it does not require special care and precisely because of this is frequently under diagnosed.

Okadaic acid is a highly soluble in lipids polyether, thermoresistant and with an irritating and cytotoxic action. This is why it persists in preserved or cooked shellfish. Furthermore, it is a strong tumour promoter. It is unknown the long term effect of the toxin capable of being retained in the body fat, although there is some evidence about its effect as a risk factor in the colon cancer[12, 14].

In Europe, the amount considered safe to prevent DSP is 0.16 grams per gram of edible portion of the mollusk[37]. However, these limits are often surpassed in other parts of the globe, and in countries with control systems in place, like Spain, the detected levels can be close to the limit. The limit is established because is the one in the mouse bioassay; below this level it could only be detected by means of more sensible procedures like ELISA or HPLC. The Bioassay must continue to be the diagnostic reference to detect presence of toxins in molluscs because it is very unspecific; it will detect any new toxin without loosing sensibility, especially if other assays are performed to confirm and to be more precise in detecting toxins.

Control systems are more efficient if they focus in monitoring microorganisms that produce toxins. The absence of the toxic dinoflagellate implies the absence of toxin. Control systems based in detecting toxins must consider that some of these (the liposoluble ones) can have long-term toxic effect developed only in sensitive people due to hereditary factors and in foetus and breast-feeding babies (certainly is excreted this via). Although this is so far only a theoretical possibility, the best safe level for DSP toxins is the absence of toxin, and this can be guaranteed with good control systems.

Global Change, Marine Aquacultures and Biotoxins

The argument that extractive way of life must be necessarily changed to another mainly based on intensive fish farming is basically false, because the farmed fish and shellfish need primarily marine foodstuffs originated from extractive systems. Further more, extractive systems have an impact on the environment not only because of the freedom of specimens but also because of pollutants released into the ocean. Extractive fishery is environmentally sustainable if done in a regulated way and knowing the quantities present of the commercial species and adapting to them, with the biological stops that are recommended.

Intensive mussel culture in *batea* is a production that has to progress with special attention to security and quality. For so, care and respect to the environment has to be optimum. Increasingly contaminated estuaries cannot produce better quality molluscs although they produce more quantity. Safer products reinforce the consumer security and increase the added value of the product. Excessive increases in productions always affect prices negatively.

Water eutrofization due to rises in water temperature and to spillages rich in phosphates certainly increase red tides, not only in means of occurrence, but also in duration and intensity, with an increase in the days that extraction is banned in the shellfish producer regions.

These foodstuffs are natural or antropogenic toxin and metal concentrators, and its consumption must be promoted as something occasional, not on a daily basis or as a basic nutrient.

References

[1] USDA National Nutrient Database for Standard Reference. Available from: http://www.nal.usda.gov/fnic/foodcomp/search/

[2] Yoshimura T. Yusho in Japan. Ind Health. 2003 Jul;41(3):139-48.

[3] Nakagawa R, Hirakawa H, Hori T. Estimation of 1992-1993 dietary intake of organochlorine and organophosphorus pesticides in Fukuoka, Japan.J AOAC Int. 1995 Jul-Aug;78(4):921-9.

[4] Sioen I, De Henauw S, Van Camp J. Evaluation of benefits and risks related to seafood consumption. Verh K Acad Geneeskd Belg. 2007;69(5-6):249-89.

[5] Wang WX, Chow AT.Benzo[a]pyrene absorption and exposure pathways in the green mussel *Perna viridis*. *Environ Toxicol Chem.* 2002 Feb;21(2):451-8.

[6] Morales-Caselles C, Ramos J, Riba I, Delvalls TA. Using the polychaete *Arenicola marina* to determine toxicity and bioaccumulation of PAHS bound to sediments. Environ Monit Assess. 2007 Sep 18 [Epub ahead of print].

[7] Shi D, Wang WX. Uptake of aqueous and dietary metals by mussel *Perna viridis* with different Cd exposure histories. *Environ Sci Technol.* 2005 Dec 1;39(23):9363-9.

[8] Ekino S, Susa M, Ninomiya T, Imamura K, Kitamura T.Minamata disease revisited: an update on the acute and chronic manifestations of methyl mercury poisoning. *J. Neurol. Sci.* 2007 Nov 15;262(1-2):131-44.

[9] Faustman EM, Ponce RA, Ou YC, Mendoza MA, Lewandowski T, Kavanagh T. Investigations of methylmercury-induced alterations in neurogenesis. Environ Health Perspect. 2002 Oct;110 Suppl 5:859-64.

[10] Hernández C, Ulloa J, Vergara JA, Espejo R, Cabello F. Vibrio parahaemolyticus infections and algal intoxications as emergent public health problems in Chilenish Rev. Med. Chil. 2005 Sep;133(9):1081-8.

[11] Gomez-Bautista M, Ortega-Mora LM, Tabares E, Lopez-Rodas V, Costas E.Detection of infectious Cryptosporidium parvum oocysts in mussels (Mytilus galloprovincialis) and cockles (Cerastoderma edule). *Appl. Environ. Microbiol.* 2000 May;66(5):1866-70.

[12] Hwang DF, Noguchi T. Tetrodotoxin poisoning. Adv Food Nutr Res. 2007;52:141-236.

[13] Passage of Darkness: The Ethnobiology of the Haitian Zombie Wade Davis. University of North Carolina Press (31 Dec 1988).

[14] US FDA.Center for Food Safety and Applied Nutrition. Foodborne Pathogenic Microorganisms and Natural Toxins Handbook. Scombrotoxin. Available from: http://www.cfsan.fda.gov/~mow/chap38.html

[15] López-Rodas V, Maneiro E, Martínez J, Navarro M, Costas E. Harmful algal blooms, red tides and human health: Diarrhetic shelfish poisoning and colorrectal cancer. Anales de la Real Academia Nacional de Farmacia. 2006, 3:391-408.

[16] Manerio E, Rodas VL, Costas E, Hernandez JM. Shellfish consumption: a major risk factor for colorectal cancer. Med Hypotheses. 2008;70(2):409-12. Epub 2007 Jul 2.

[17] Cordier S, Monfort C, Miossec L, Richardson S, Belin C. Ecological analysis of digestive cancer mortality related to contamination by diarrhetic shellfish poisoning toxins along the coasts of France. *Environ. Res.* 2000 Oct;84(2):145-50.

[18] Paralytic Shellfish Poisoning -- Massachusetts and Alaska, 1990 MMWR March 15, 1991 / 40(10);157-161.

[19] Rodrigue DC, Etzel RA, Hall S, de Porras E, Velasquez OH, Tauxe RV, Kilbourne EM, Blake PA. Lethal paralytic shellfish poisoning in Guatemala. *Am. J. Trop Med. Hyg.* 1990 Mar;42(3):267-71.

[20] La Barbera-Sánchez A, Franco Soler J, Rojas de Astudillo L, Chang-Yen I. Paralytic shellfish poisoning (PSP) in Margarita Island, Venezuela. Rev Biol Trop. 2004 Sep;52 Suppl 1:89-98.

[21] Gago-Martinez A, Rodriguez-Vazquez JA, Thibault P, Quilliam MA. Simultaneous occurrence of diarrhetic and paralytic shellfish poisoning toxins in Spanish mussels in 1993. *Nat. Toxins.* 1996;4(2):72-9.

[22] Wu JY, Zheng L, Wang JH. Contamination of shellfish from Shanghai seafood markets with paralytic shellfish poisoning and diarrhetic shellfish poisoning toxins determined by mouse bioassay and HPLC. *Food Addit Contam.* 2005 Jul;22(7):647-51.

[23] Jakobsen B, Tasker A, Zimmer J.Domoic acid neurotoxicity in hippocampal slice cultures. Amino Acids. 2002;23(1-3):37-44.

[24] Jeffery B, Barlow T, Moizer K, Paul S, Boyle C. Amnesic shellfish poison. *Food Chem. Toxicol.* 2004 Apr;42(4):545-57

[25] Hess P, Gallacher S, Bates LA, Brown N, Quilliam MA. Determination and confirmation of the amnesic shellfish poisoning toxin, domoic acid, in shellfish from Scotland by liquid chromatography and mass spectrometry.J AOAC Int. 2001 Sep-Oct;84(5):1657-67.

[26] Amzil Z, Fresnel J, Le Gal D, Billard C.Domoic acid accumulation in French shellfish in relation to toxic species of Pseudo-nitzschia multiseries and P. pseudodelicatissima. *Toxicon.* 2001 Aug;39(8):1245-51.

[27] James KJ, Gillman M, Amandi MF, López-Rivera A, Puente PF, Lehane M, Mitrovic S. Furey A.Amnesic shellfish poisoning toxins in bivalve molluscs in Ireland. *Toxicon.* 2005 Dec 15;46(8):852-8.

[28] Stobo LA, Lacaze JP, Scott AC, Petrie J, Turrell EA. Surveillance of algal toxins in shellfish from Scottish waters. *Toxicon.* 2008 Mar 15;51(4):635-48.

[29] Costa PR, Rosa R, Duarte-Silva A, Brotas V, Sampayo MA. Accumulation, transformation and tissue distribution of domoic acid, the amnesic shellfish poisoning toxin, in the common cuttlefish, Sepia officinalis. *Aquat Toxicol.* 2005 Aug 15;74(1):82-91.

[30] Lefebvre KA, Bargu S, Kieckhefer T, Silver MW. From sanddabs to blue whales: the pervasiveness of domoic acid. *Toxicon.* 2002 Jul;40(7):971-7.

[31] Council Directive 97/61/EC of 20 October 1997 amending the Annex to Directive 91/492/EEC laying down the health conditions for the production and placing on the market of live bivalve molluscs.

[32] Programa de sanidad de moluscos bivalvos. Norma Técnica Sección 3. Servicio Nacional de Pesca Gobierno de Chile.. 2006. Available from: http://www2.sernapesca.cl/lib /doc_atributo.php?c=001007008001007002003

[33] Backer LC, Kirkpatrick B, Fleming LE, Cheng YS, Pierce R, et al. Occupational exposure to aerosolized brevetoxins during Florida red tide events: effects on a healthy worker population. *Environ Health Perspect.* 2005 May;113(5):644-9.

[34] Milian A, Nierenberg K, Fleming LE, Bean JA, Wanner A, Reich A, Backer LC, Jayroe D, Kirkpatrick B. Reported respiratory symptom intensity in asthmatics during exposure to aerosolized Florida red tide toxins. *J. Asthma.* 2007 Sep;44(7):583-7.

[35] Durando P, Ansaldi F, Oreste P, Moscatelli P, Marensi L, Grillo C, Gasparini R, Icardi G. Ostreopsis ovata and human health: epidemiological and clinical features of respiratory syndrome outbreaks from a two-year syndromic surveillance, 2005-06, in north-west Italy. Euro Surveill. 2007;12(23):pii=3212. Available online: http://www. eurosurveillance.org /ViewArticle.aspx?ArticleId=3212

[36] Onuma Y, Satake M, Ukena T, Roux J, Chanteau S, Rasolofonirina N, et al. Identification of putative palytoxin as the cause of clupeotoxism. Toxin 1999; 37: 55-65.

[37] Commission Decision of 15 March 2002 laying down detailed rules for the implementation of Council Directive 91/492/EEC as regards the maximum levels and the methods of analysis of certain marine biotoxins in bivalve molluscs, echinoderms, tunicates and marine gastropods. *Official Journal of the European Communities.* (2002/225/EC) L 75/62-63.

In: Fish Consumption and Health
Editors: George P. Gagne and Richard H. Medrano
ISBN 978-1-60741-151-2

Chapter X

Role of Parenteral Fish Oil Lipid Emulsions in Pediatric Nutrition: Benefits and Concerns

Kathleen M. Gura*
Department of Pharmacy
Children's Hospital Boston, Massachusetts, USA

Abstract

Until recently, all patients receiving parenteral nutrition (PN) received all or a portion of their fat calories from lipid emulsions derived from soybean oils. It was felt that using another fat source alone, such as fish oil, would predispose patients to developing essential fatty acid deficiency (EFAD), bleeding, or other complications. This review describes recent developments in the use of fish oil based intravenous fat emulsions as monotherapy in the PN dependent patient as it relates to the reversal of PN associated liver disease and provides an overview of why complications such as bleeding or EFAD are rare. Other benefits of using fish oil as the preferred parenteral lipid source, such as prevention of retinopathy of prematurity and hypertriglyceridemia, will also be discussed.

Introduction

Parenteral nutrition (PN) has been an established modality providing an necessary energy and nutrients for patients unable to absorb enteral feeding due to either intestinal abnormalities, or when the use of gastrointestinal tract is contraindicated.[1] In surgical

* Corresponding Author: Kathleen M. Gura, Pharm.D.Postal address: Department of Pharmacy, Children's Hospital Boston 300 Longwood Avenue Boston, 02115, USA, Phone: 617-355- 2336 Fax: 617-730-0601, E-mail: Kathleen.Gura@childrens.harvard.edu.

patients with short bowel syndrome, PN serves as an the initial mainstay of therapy has become integral part of the care for premature infants. In thc United States more than 30,000 patients are permanently dependent on PN for survival.[2]

Long-term use of PN, however, is associated with many complications, including blood stream infections, metabolic abnormalities, and organ dysfunction.[3]First documented in 1971, the most prevalent and most severe complication continues to be hepatobiliary dysfunction, referred to as PN-associated liver disease (PNALD).[4] Symptoms include steatosis, cholestasis, cholelithiasis, hepatic fibrosis, that can ultimately progress to biliary cirrhosis, portal hypertension, and end-stage liver failure.[5] The reported prevalence rate of PNALD varies greatly, but it is estimated to be up to 27% in children and 85% in neonates who are on long-term PN for intestinal failure. Approximately 15% of patients who receive long-term PN eventually develop end-stage liver disease.[6-8] Until recently, the only effective treatment for patients with intestinal failure who have progressed to end-stage liver disease is isolated, or combined, liver-small bowel transplantation, which is, unfortunately, associated with a high mortality itself.[9]

The etiology of PNALD is complex and multi-factorial. Risk factors for the development of PNALD include: prematurity, low birth weight, duration of PN and number of septic events. [5, 8, 10, 11] Increasing enteral caloric intake while reducing PN is the most effective treatment for PNALD, but this process may be impossible when intestinal function is poor. Recently evidences suggests that the composition of the intravenous lipid emulsion, which is typically administered in combination with PN to provide dense calories and prevent essential fatty acid deficiency (EFAD), can significantly contribute to PNALD.[12]

The first safe and effective parenteral lipid emulsion, Intralipid® (Fresenius Kabi, Bad Homburg Germany), became available in Europe in 1961.[13] This soy bean derived lipid emulsion with egg yolk phospholipids as an emulsifier has since been considered the gold standard to which other lipid emulsions are compared. It was not until the 1970's that it available in the United States. The first alternative lipid emulsion to Intralipid® that became available in the United States was Liposyn I® (Abbott Laboratories, North Chicago, IL, USA), that was derived solely from safflower oils. It was withdrawn from the market when neurological abnormalities and other adverse side effects were noted due to its low linolenic acid (omega-3) content (0.1%) which negatively impacted essential fatty acid, triglyceride, and free fatty acid levels. .[14] It has since been replaced with a 50/50 blend of safflower and soybean oils (Liposyn II®), leaving American practitioners with only 2 types of emulsions, both of which contain soybean oils. (Table 1). The fear of developing EFAD most likely explains why soy bean oil continues to dominate the market and is used in all commercially available parenteral lipid emulsions suitable for administration as monotherapy worldwide.

Soy Bean Oil and PNALD

Soy bean oil, despite its widespread use as a parenteral lipid source, is not ideal. Depending on their route of administration, lipids are metabolized differently. [12] When taken orally, lipids are absorbed by the enterocyte in the form of a micelle and packaged into chylomicrons for ultimate disposal in the liver.

Table 1. Comparison of parenteral lipid emulsions per 10g/100mL

Product		Intralipid®	Liposyn II®	Omegaven®
Manufacturer		Fresenius Kabi*	Hospira	Fresenius Kabi
Oil source (%)				
Soy bean		100	50	–
Safflower		–	50	–
Fish		–	–	100
Phytosterols (mg/L)		348±33	383	–
Fat composition (g)				
Palmitic	(16:0)	1	0.88	0.25-1
Stearic	(18:0)	0.4	0.32	0.05-0.2
Oleic	(18:1)	2.6	1.82	0.6-1.3
Linoleic	(18:2)	5	6.54	0.1-0.7
α-Linolenic	(18:3)	0.9	0.39	<0.2
EPA	(20:5)	–	–	1.28-2.82
Arachidonic	(20:4)	–	–	0.1 -0.4
DHA	(22:5)	–	–	1.44-3.09

Data as provided by manufacturers
EPA: eicosapentaenoic acid; DHA: docosahexaenoic acid.

The emulsified particles of commercially made and intravenously administered lipid emulsions, such as Intralipid®, mimic the size and structure of chylomicrons, but differ in their content.[15] In contrast to chylomicrons, these artificial lipid particles primarily contain essential fatty acids and omega-6 triglycerides, but are devoid of cholesterol or protein. Recent studies suggest they may be cleared as whole particles by tissues other than the liver which may account for some of the complications associated with their use.[16] Moreover, phytosterols found in soy bean oils are thought to have a deleterious effect on hepatic function.[17,18] In one animal study, intravenous phytosterol injections without the other components of parenteral lipid emulsions markedly reduced bile acid excretion.[19] Furthermore, long-term use of a soy bean-based lipid emulsion has been shown to cause an increase and accumulation of phytosterol content in cell membranes and plasma lipoproteins, which was has been linked with the onset of cholestasis in children on prolonged courses of PN.[17]

Another key point that suggests that soybean emulsions may contribute to the development and progression of PNALD is inflammation. Soybean oils are rich in the omega-6 fatty acids linoleic acid (LA) and arachidonic acid (AA), which can increase the production of inflammatory mediators such as tumor necrosis factor alpha and interleukin-6. Furthermore, AA is a key substrate for the potent pro-inflammatory mediators, including eicosanoids prostaglandin E2, leukotriene B4, thromboxane 2 and platelet aggregation factor.[20] It has been suggested that increased levels of both AA and LA can initiate or worsen inflammatory states.[21] Moreover, *in vitro* and clinical studies have demonstrated that soy bean-based lipid emulsions have immunosuppressive effects [22,23] These factors could explain the link between PNALD with parenteral administration of soy bean-based lipid emulsions.

Fish Oil as an Alternative Lipid Source

Fish oil-based lipid emulsions as monotherapy has emerged as an exciting alternative to soy bean oil based lipid emulsions. These products have been available in Europe and Asia for more than 10 years but have been marketed as a *supplement* to be infused with conventional soy bean-based lipid emulsions. Due to its high concentration of eicosapentaenoic acid (EPA, omega-3) and docosahexaenoic acid (DHA, omega-3), fish oil has been known to have anti-inflammatory potential by interfering with the AA pathway and producing the anti-inflammatory eicosanoids prostaglandin E3, leukotriene B5 and thromboxane 3.[24-26] The use of fish oil in lipid emulsions for PN alone was initially hindered by skepticism that fish oil might not provide sufficient essential fatty acids to prevent EFAD and to enhance growth.[27]

Fish oil modulates inflammation through the inhibition of tumor necrosis factor-alpha and by the reduction of cytokines that trigger pro-inflammatory reactions.[28-31] Using a rodent model, Yeh et al. compared groups of rats receiving PN with either safflower oil based or a fish oil-based lipid emulsions.[32] They theorized that omega-3 fatty acids may play a role in reducing radicals that contribute to the suppression of the inflammatory pathway as demonstrated by reduced superoxide dismutase and glutathione levels in rats fed fish oil. This is consistent with several other animal studies showing that parenteral fish oil supplementation with PN showed a marked reduction in bile flow and improvement in PN-associated cholestasis.[33,34] Moreover, omega-3 fatty acids have been shown to accelerate the clearance of triglycerides in humans, thus effectively reducing triacylglycerol concentrations in serum in vivo.[35] Others have shown through suppression of sterol response element binding protein-1, an increase in clearance of triglycerides and a reduction in lipogenesis occur that may provide a clue as to as why omega 3 fatty acids may be less hepatotoxic in the PN dependent patient.[36,37]

Based on these aforementioned findings, it has been hypothesized that fish oil supplementation could attenuate fatty liver changes in a murine model of PNALD.[38,39] Mice were fed exclusively with an *ad libitum* liquid fat-free diet containing 20% dextrose, similar to the PN used in patients. On this fat free diet, the animals developed PNALD in addition to EFAD and enhanced *de novo* lipogenesis. [39] To test the hypothesis that fish oil supplementation could prevent PNALD, groups of mice received PN with either a conventional soy bean oil-based emulsion or a fish oil-based emulsion. Control mice received PN without lipid supplementation. Severe steatosis was observed in animals that received PN alone or PN supplemented with the standard soy bean-based lipid emulsion. This was characterized by diffuse macro- and microvesicular steatosis on histology, as well as elevated serum alanine aminotransferase and aspartase aminotransferase levels. Mice given fish oil supplements did not develop steatosis, and had normal hepatic histology and normal liver function tests. EFAD was not observed in either the soy or fish oil groups. [39]

Clinical Experience with Fish Oil Based Lipid Emulsions

Despite evidence to the contrary in a variety of animal models, there is still skepticism regarding the ability of parenteral fish oil to prevent EFAD and maintain adequate growth rate when used as monotherapy. Recent reports suggest that the omega-3 fatty acids, alpha-linolenic acid (ALA), EPA, and DHA, which are present in fish oil, might play a role in reducing the amount of the omega-6 fatty acid, linoleic acid required to prevent EFAD.[40]

Omegaven® (Fresenius Kabi Deutschland GmbH, Bad Homburg v.d.H., Germany) is the only commercially available lipid emulsion that contains solely fish oils. In 2005, Gura *et al.* described the first case in which Omegaven® was used as the sole intravenous lipid source in a human.41 Due to a severe soy allergy, this patient was unable to tolerate any parenteral fat emulsion commercially available in the United States. The patient had gone to develop biochemical evidence of EFAD accompanied by truncal folliculitis over a course of months secondary to fat free PN in the absence of any enteral feedings and despite topical oil therapy. [41] After receiving emergency FDA approval, Omegaven® was administered and resulted in resolution of the rash in 10 days and normalization of essential fatty acid status after 17 days.

Since this case report and based on the results of the previously described murine studies, Omegaven® monotherapy has been used Children's Hospital Boston since 2004 under an FDA approved compassionate treatment protocol for PN-dependent patients with cholestasis. In one early case report, 2 pediatric patients with cholestasis due to PNALD were treated with Omegaven® and their serum bilirubin levels normalized within 60 days. [42] Both infants tolerated the therapy well and did not experience any adverse reactions. Based on these initial findings, more than 100 children with PNALD with the fish oil-based lipid emulsion have been similarly treated. In March 2008, a comprehensive analysis was published in which the outcome of the first 18 patients on Omegaven® monotherapy [43] was compared to a group of 21 historical controls who received the conventional soybean oil-based lipid emulsions.[11] Children who received Omegaven® monotherapy experienced reversal of cholestasis approximately 7 times faster than the historical controls, effectively halting the progression of PNALD. This gave the children time to develop enteral tolerance and transition to full enteral feedings. In this study, 2 children receiving Omegaven® died from causes unrelated to its use, whereas mortality related to PNALD was 29% in the historical cohort. Moreover, 2 infants receiving parenteral soy bean-based lipid emulsions underwent a liver transplantation, compared to no liver transplants in the Omegaven® oil treated patients. Furthermore, Omegaven® monotherapy was not associated with EFAD, hypertriglyceridemia, coagulopathy, bloodstream infections, or growth impairment. Since the publication of these reports, several other centers have experience similar dramatic findings. [44, 45]

Additional Benefits of Fish Oil

It was been well known that neonatal development is dependent of long chain fatty essential fatty acids. Until recently, infants given enteral formulas not supplemented with DHA or AA were at risk for clinical deficiencies that were evident in the form of growth impairment dermatologic abnormalities (dryness, desquamation, thickening), abnormal visual function, peripheral neuropathy, and neurodevelopment deficiencies. [46] The development of and function of neural tissue is dependent on long chain fatty acids which impact both cell membrane composition and fluidity. In the retina, the rod photoreceptor has membranes composed of 30%-40% DHA. [47] Recent work by Connor and colleagues suggests supplementing omega-3-fatty acid intake may be of benefit in preventing retinopathy of prematurity, one of the leading causes of childhood blindness.[48] By providing neonates a lipid emulsion containing fish oils, in addition to being less hepatotoxic, such products may offer additional neurodevelopmental benefits not present in current soy containing emulsions.

Conclusion

Mounting evidence indicates that these plant oil-based lipid emulsions significantly contribute to PNALD. Recent studies have turned towards fish oil-based lipid emulsions, because of their high omega-3 fatty acid content, to determine if they are indeed hepatoprotective. The previously described experimental studies have shown that supplementation of fish oil markedly prevented and reversed PNALD without the development of EFAD. Moreover, recently published clinical studies have shown that an intravenous fish oil-based lipid emulsion as monotherapy in pediatric patients dependent on PN is a safe and efficacious method of reversing PNALD and normalizing EFAD status. Additional research in this area is still warranted and prospective randomized clinical trials are still needed to fully understand the effectiveness of this new treatment strategy.

References

[1] Dudrick SJ, Wilmore DW, Vars HM, Rhoads JE. Long-term total parenteral nutrition with growth, development, and positive nitrogen balance. *Surgery* 1968;64:134-142.

[2] Howard L, Ament M, Fleming CR, Shike M, Steiger E. Current use and clinical outcome of home parenteral and enteral nutrition therapies in the United States. *Gastroenterology* 1995;109:355-365.

[3] Carter BA, Karpen SJ. Intestinal failure-associated liver disease: management and treatment strategies past, present, and future. *Semin. Liver Dis.* 2007;27:251-258.

[4] Peden VH, Witzleben CL, Skelton MA. Total parenteral nutrition. *J. Pediatr.* 1971;78:180-181.

[5] Kelly DA. Liver complications of pediatric parenteral nutrition--epidemiology. *Nutrition* 1998;14:153-157.

[6] Diamanti A, Basso MS, Castro M, Calce A, Pietrobattista A, Gambarara M. Prevalence of life-threatening complications in pediatric patients affected by intestinal failure. *Transplant Proc.* 2007;39:1632-1633.

[7] Chan S, McCowen KC, Bistrian BR, Thibault A, Keane-Ellison M, Forse RA, Babineau T, et al. Incidence, prognosis, and etiology of end-stage liver disease in patients receiving home total parenteral nutrition. *Surgery* 1999;126:28-34.

[8] Christensen RD, Henry E, Wiedmeier SE, Burnett J, Lambert DK. Identifying patients, on the first day of life, at high-risk of developing parenteral nutrition-associated liver disease. *J.Perinatol.* 2007;27:284-290.

[9] Nathan JD, Rudolph JA, Kocoshis SA, Alonso MH, Ryckman FC, Tiao GM. Isolated liver and multivisceral transplantation for total parenteral nutrition-related end-stage liver disease. *J. Pediatr. Surg.*2007;42:143-147.

[10] Teitelbaum DH, Tracy T. Parenteral nutrition-associated cholestasis. *Semin. Pediatr. Surg.* 2001;10:72-80.

[11] Andorsky DJ, Lund DP, Lillehei CW, Jaksic T, Dicanzio J, Richardson DS, Collier SB, et al. Nutritional and other postoperative management of neonates with short bowel syndrome correlates with clinical outcomes. *J. Pediatr.* 2001;139:27-33.

[12] Javid PJ, Greene AK, Garza J, Gura K, Alwayn IP, Voss S, Nose V, et al. The route of lipid administration affects parenteral nutrition-induced hepatic steatosis in a mouse model. *J. Pediatr. Surg*.2005;40:1446-1453.

[13] Schuberth O, Wretlind A. Intravenous infusion of fat emulsions, phosphatides and emulsifying agents. *Acta Chir .Scand*.1961;13(suppl):S278-284.

[14] Holman RT, Johnson SB, Hatch TF. A case of human linolenic acid deficiency involving neurological abnormalities. *Am. J. Clin. Nutr.* 1982;35:617-623.

[15] Ferezou J, Bach AC. Structure and metabolic fate of triacylglycerol- and phospholipid-rich particles of commercial parenteral fat emulsions. *Nutrition* 1999;15:44-50.

[16] Qi K, Al-Haideri M, Seo T, Carpentier YA, Deckelbaum RJ. Effects of particle size on blood clearance and tissue uptake of lipid emulsions with different triglyceride compositions. *JPEN J Parenter Enteral. Nutr.* 2003;27:58-64.

[17] Clayton PT, Bowron A, Mills KA, Massoud A, Casteels M, Milla PJ. Phytosterolemia in children with parenteral nutrition-associated cholestatic liver disease. *Gastroenterology* 1993;105:1806-1813.

[18] Clayton PT, Whitfield P, Iyer K. The role of phytosterols in the pathogenesis of liver complications of pediatric parenteral nutrition. *Nutrition* 1998;14:158-164.

[19] Iyer KR, Spitz L, Clayton P. BAPS prize lecture: New insight into mechanisms of parenteral nutrition-associated cholestasis: role of plant sterols. British Association of Paediatric Surgeons. *J. Pediatr. Surg.* 1998;33:1-6.

[20] Tilley SL, Coffman TM, Koller BH. Mixed messages: modulation of inflammation and immune responses by prostaglandins and thromboxanes. *J. Clin. Invest* 2001;108:15-23.

[21] Calder PC. Use of fish oil in parenteral nutrition: Rationale and reality. *Proc. Nutr. Soc.* 2006;65:264-277.

[22] Calder PC, Sherrington EJ, Askanazi J, Newsholme EA. Inhibition of lymphocyte proliferation in vitro by two lipid emulsions with different fatty acid compositions. *Clin. Nutr.* 1994;13:69-74.

[23] Monson JR, Sedman PC, Ramsden CW, Brennan TG, Guillou PJ. Total parenteral nutrition adversely influences tumour-directed cellular cytotoxic responses in patients with gastrointestinal cancer. *Eur. J. Surg .Oncol.* 1988;14:935-943.

[24] Chen MF, Lee YT, Hsu HC, Yeh PC, Liau CS, Huang PC. Effects of dietary supplementation with fish oil on prostanoid metabolism during acute coronary occlusion with or without reperfusion in diet-induced hypercholesterolemic rabbits. *Int. J. Cardiol.* 1992;36:297-304.

[25] Lee S, Gura KM, Kim S, Arsenault DA, Bistrian BR, Puder M. Current clinical applications of omega-6 and omega-3 fatty acids. Nutr Clin Pract 2006;21:323-341.

[26] Lee S, Gura KM, Puder M. Omega-3 fatty acids and liver disease. *Hepatology* 2007;45:841-845.

[27] Carlson SE, Cooke RJ, Werkman SH, Tolley EA. First year growth of preterm infants fed standard compared to marine oil n-3 supplemented formula. Lipids 1992;27:901-907.

[28] Babcock TA, Novak T, Ong E, Jho DH, Helton WS, Espat NJ. Modulation of lipopolysaccharide-stimulated macrophage tumor necrosis factor-alpha production by omega-3 fatty acid is associated with differential cyclooxygenase-2 protein expression and is independent of interleukin-10. *J. Surg. Res.* 2002;107:135-139.

[29] Mayer K, Meyer S, Reinholz-Muhly M, Maus U, Merfels M, Lohmeyer J, Grimminger F, et al. Short-time infusion of fish oil-based lipid emulsions, approved for parenteral nutrition, reduces monocyte proinflammatory cytokine generation and adhesive interaction with endothelium in humans. *J. Immunol.* 2003;171:4837-4843.

[30] Novak TE, Babcock TA, Jho DH, Helton WS, Espat NJ. NF-kappa B inhibition by omega -3 fatty acids modulates LPS-stimulated macrophage TNF-alpha transcription. *Am. J. Physiol. Lung Cell Mol .Physiol* .2003;284:L84-89.

[31] Rallidis LS, Paschos G, Liakos GK, Velissaridou AH, Anastasiadis G, Zampelas A. Dietary alpha-linolenic acid decreases C-reactive protein, serum amyloid A and interleukin-6 in dyslipidaemic patients. *Atherosclerosis* 2003;167:237-242.

[32] Yeh SL, Chang KY, Huang PC, Chen WJ. Effects of n-3 and n-6 fatty acids on plasma eicosanoids and liver antioxidant enzymes in rats receiving total parenteral nutrition. *Nutrition* 1997;13:32-36.

[33] Van Aerde JE, Duerksen DR, Gramlich L, Meddings JB, Chan G, Thomson AB, Clandinin MT. Intravenous fish oil emulsion attenuates total parenteral nutrition-induced cholestasis in newborn piglets. *Pediatr. Res .1*999;45:202-208.

[34] Chen WJ, Yeh SL. Effects of fish oil in parenteral nutrition. Nutrition 2003;19:275-279.

[35] Putadechakum S, Tanphaichitr V, Leelahagul P, Pakpeankitvatana V, Surapisitchart T, Komindr S. Long-term treatment of N-3 PUFAS on plasma lipoprotein levels and fatty acid composition of total serum and erythrocyte lipids in hypertriglyceridemic patients. *J. Med. Assoc. Thai* .2005;88:181-186.

[36] Park Y, Harris WS. Omega-3 fatty acid supplementation accelerates chylomicron triglyceride clearance. *J. Lipid Res.* 2003;44:455-463.

[37] Sekiya M, Yahagi N, Matsuzaka T, Najima Y, Nakakuki M, Nagai R, Ishibashi S, et al. Polyunsaturated fatty acids ameliorate hepatic steatosis in obese mice by SREBP-1 suppression. *Hepatology* 2003;38:1529-1539.

[38] Alwayn IP, Javid PJ, Gura KM, Nose V, Ollero M, Puder M. Do polyunsaturated fatty acids ameliorate hepatic steatosis in obese mice by SREPB-1 suppression or by correcting essential fatty acid deficiency. Hepatology 2004;39:1176-1177; author reply 1177-1178.

[39] Alwayn IP, Gura K, Nose V, Zausche B, Javid P, Garza J, Verbesey J, et al. Omega-3 fatty acid supplementation prevents hepatic steatosis in a murine model of nonalcoholic fatty liver disease. *Pediatr. Res.* 2005;57:445-452.

[40] Strijbosch RA, Lee S, Arsenault DA, Andersson C, Gura KM, Bistrian BR, Puder M. Fish oil prevents essential fatty acid deficiency and enhances growth: clinical and biochemical implications. *Metabolism* 2008;57:698-707.

[41] Gura KM, Parsons SK, Bechard LJ, Henderson T, Dorsey M, Phipatanakul W, Duggan C, et al. Use of a fish oil-based lipid emulsion to treat essential fatty acid deficiency in a soy allergic patient receiving parenteral nutrition. *Clin. Nutr.* 2005;24:839-847.

[42] Gura KM, Duggan CP, Collier SB, Jennings RW, Folkman J, Bistrian BR, Puder M. Reversal of parenteral nutrition-associated liver disease in two infants with short bowel syndrome using parenteral fish oil: implications for future management. *Pediatrics* 2006;118:e197-201.

[43] Gura KM, Lee S, Valim C, Zhou J, Kim S, Modi BP, Arsenault DA, et al. Safety and efficacy of a fish-oil-based fat emulsion in the treatment of parenteral nutrition-associated liver disease. *Pediatrics* 2008;121:e678-686.

[44] Diamond IR, Sterescu A, Pencharz PB, Wales PW. The rationale for the use of parenteral omega-3 lipids in children with short bowel syndrome and liver disease. *Pediatr. Surg. Int.* 2008 Jul;24(7):773-8. Epub 2008 May 27.

[45] Ekema G, Falchetti D, Boroni G, Tanca AR, Altana C, Righetti L, Ridella M, Gambarotti M, Berchich L.Reversal of severe parenteral nutrition-associated liver disease in an infant with short bowel syndrome using parenteral fish oil (Omega-3 fatty acids).*J. Pediatr .Surg.* 2008 Jun;43(6):1191-5.

[46] 46 Holman RT, Johnson SB, Hatch TF. A case of human linolenic acid deficiency involving neurological abnormalities. *Am. J. Clin. Nutr.* 1982;35:617 -623.

[47] Fliesler SJ, Anderson RE. Chemistry and metabolism of lipids in the vertebrate retina. *Prog. Lipid. Res.*1983; 22:79 -131.

[48] Connor KM, SanGiovanni JP, Lofqvist C, Aderman CM, Chen J, Higuchi A, Hong S, Pravda EA, Majchrzak S, Carper D, Hellstrom A, Kang JX, Chew EY, Salem N Jr, Serhan CN, Smith LE. Increased dietary intake of omega-3-polyunsaturated fatty acids reduces pathological retinal angiogenesis. Nat Med. 2007. Jul;13(7):868-73. Epub 2007 Jun 24.

In: Fish Consumption and Health
Editors: George P. Gagne and Richard H. Medrano
ISBN 978-1-60741-151-2

Chapter XI

Do Specific Immunoglobulin E Levels Have any Role in the Diagnostic of Corn Allergy in Children?

Elizabeth Estrada-Reyes,[1] Mirna Toledo-Bahena,[2] and Alejandro A. Nava-Ocampo[3,4]
[1]Consulta de Alergología Pediátrica
[2]Dermatología Pediátrica, Hospital Ángeles Metropolitano, Mexico DF, México
[3]Department of Pharmacology and Toxicology, Faculty of Medicine, University of Toronto, Toronto, Canada
[4]PharmaReasons, Toronto, Canada

Background

Cereals, of the taxonomic family Gramineae, are responsible for a high rate of allergic and immunologic reactions [1,2]. Cereals include wheat, barley, rye, oats, rice and maize, and different consumption patterns of various cereals around the world may justify the different rates of sensitization reported for each cereal [1]. They are the major components of the human diet in many countries around the world, and sensitization to any of the cereals consequently represents a major public health challenge. In many countries, corn (Zea mays L) is the basic ingredient of frequently eaten foods like porridge, polenta, corn flakes, popcorn, tacos, chips, canned de-embryonated kernels, and other food products. Corn is also component of products used in the elaboration of bread, cakes and beer. In addition, corn may be used in the elaboration of starch, glue, cosmetics, stickers, stamp adhesives, exipients (for example, in the formulations of aspirin and vitamins), chewing gum, talcum powder, lozenges, ointments, toothpastes, laxatives and bath powder [3]

The protein composition of maize seeds has been investigated in depth. However, few studies have addressed the identification of maize allergens. A major lipid transfer protein

(LTP) allergen of maize appears to be a protein belonging to group 1 of grass pollen allergens that are responsible for seasonal respiratory symptoms [4]; the maize allergen appears to be responsible for provoking food-induced allergen reactions. Similar to flour, particles of corn can also be inhaled by people handling it while making cakes, pastry cooks, by housewives and by farmers [4]. Corn flour contains about 7% to 13% protein that can be fractionated into several solubility classes; the salt-extractable fraction (albumins and globulins) averages less than 4% of the total endosperm protein and mainly comprises of proteins with metabolic functions whereas 60-70% of proteins are called zeins [5].

Corn is often cited as an allergenic food and positive specific IgE to this cereal have been reported. In Mexico, despite the high consumption pattern of corn, there is only limited information on food-induced allergic reactions to maize in adults [6]. Symptoms may include urticaria, diarrhea, wheezing, and in some cases anaphylaxis [7]. Therefore, we performed the present study in order to evaluate specific IgE levels as a diagnostic test of corn allergy in children.

Patients and Methods

In a prospective study design, 41 consecutive pediatric patients between the age of 9 months and 11.5 years (median 2.3), with history of urticaria, atopic dermatitis, vomiting, diarrhea, abdominal pain, rhinitis, or wheezing after ingesting corn, were included in the study. Specific IgE levels were obtained from all children and food oral challenges to maize was performed as follows. Blood samples were obtained before the oral challenge to maize and specific IgE levels to corn were obtained by means of the Pharmacia CAP System FEIA (Pharmacia Diagnostics AB, Uppsala, Sweden).

Food challenges to maize were performed in adherence to the EAACI guidelines and following protocols reported elsewhere [8-11]. Starting from a dosage of 0.5 g, food dose was double every 20 minutes until the entire meal was eaten or until symptoms appeared. The maximum dosage was 100 g. The sensitivity, specificity, and positive and negative predictive values (PPV and NPV, respectively) were obtained for specific IgE were obtained using the food challenge as a gold standard. Receiver-operator characteristics (ROC) curves were also constructed. Area under the curve (AUC) was calculated to quantify the accuracy of the single test and to compare the diagnostic value.

Results

There were 25 boys (60.9%) and 16 were girls (39.1%). The diagnostic challenge test positive in 29 children (70%). Of them, 29 (70%) had wheezing, 16 (39%) atopic dermatitis (10 points of SCORAD), 10 (24%) diarrhea, and 8 (19.5%) urticaria. In the analysis of the ROC curve, the optimal decision point of IgE was 7.5 Ku/L with a negative predictive value of 80%, positive predictive value 66%, sensitivity 96%, and specificity 77%.

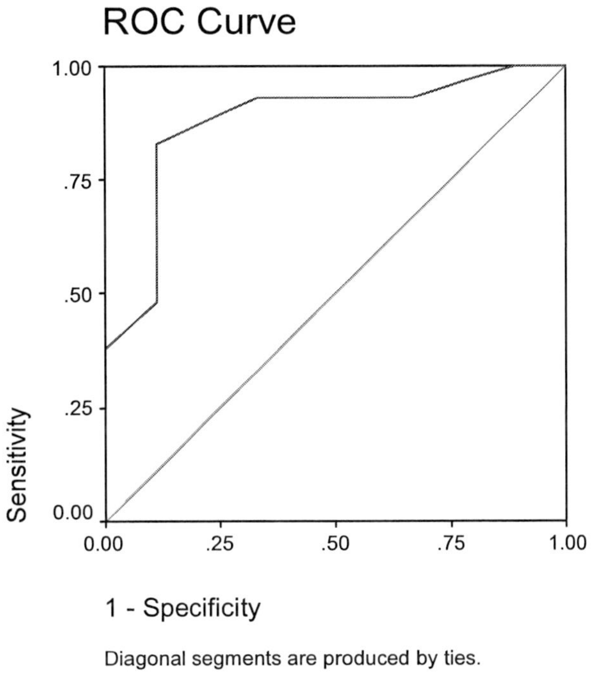

Cut-off point	Specificity (%)	Sensitivity (%)	NPV (%)	PPV (%)
7.5	77	96	80	66

Discussion

Although a definitive diagnosis can only be established by an oral food challenge, we found that specific IgE to corn may support the presence of food allergy to maize in children. Although several cut-off points for milk, egg, soy and wheat have been published elsewhere [12-13], we were not able to find any reference value for corn in children or in adults. In this context, this is the first study analyzing the IgE levels to corn in children. However, because of the limitations related to our small sample size, it is important to accumulate more experience in order to confirm the prelimar evidence shown herein. In addition, the evaluation on whether the cut-off IgE level to corn proposed in our study may be useful to document tolerance in children with allergy to maize following a treatment period with an exclusion diet, is warranted in further studies.

References

[1] Pastorello EA, Pompei C, Pravettoni V, Farioli L, Calamari AM, Scibilia J, Robino AM, Conti A. et al Lipid-transfer protein is the major maize allergen maintaining IgE-binding activity after cooking at 100°C, as demonstrated in anaphylactic patients and

patients with positive double-blind, placebo-controlled food challenge results. J Allergy Clin Immunol 2003;112:775-783.

[2] May KL. Allergy to cereals and dairy products in adult, uncomplicated asthma: an epidemiological survey. Allergol Immunopathol (Madr). 1980;8:643-650.

[3] Texas Children's Hospital. Corn allergy. 2004; http://www.texaschildrens.org /parents/tipsarticles/acrobatdisplay.aspx?fid=344 (accessed on June 18, 2008).

[4] Pastorello EA, Farioli L, Pravettoni V, Ispano M, Scibola E, Trambaiolli C, Giuffrida MG, et al. The maize major allergen, which is responsable for food-induced allergic reactions, is a lipid transfer protein. J Allergy Clin Immunol 2000;106:744-751.

[5] Pasini G, Simonato B, Curioni A, Vincenzi S, Cristaudo A, Santucci B, Peruffo AD, Giannattasio M. IgE-mediated allergy to corn: a 50kDa protein, belonging to the Reduced Soluble Proteins, is a major allergen. Allergy 2002;57:98-106.

[6] Valencia Zavala MP, Vega Robledo GB, Sánchez Olivas MA, Duarte Diaz RJ, Oviedo CL. Maize (Zea mays): allergen or toleragen? Participation of the cereal in allergic disease and positivity incidence in cutaneous tests. Rev Alerg Mex. 2006;53:207-211.

[7] Weichel M, Vergoossen NJ, Bonomi S, Scibilia J, Ortolani C, Ballmer-Weber BK, Pastorello EA. Screening the allergenic repertoires of wheat and maize with sera from double-blind, placebo-controlled food challenge positive patients. Allergy 2006; 61:128-135.

[8] Tanaka LG, El-Dahr JM, Lehrer SB. Double-blind, placebo-controlled corn challenge resulting in anaphylaxis. J Allergy Clin Immunol 2001;107:744.

[9] Sutton SA, Assa'ad AH, Steinmetz C, Rothenberg ME. A negative, double-blind, placebo-controlled challenge to genetically modified corn. J Allergy Clin Immunol 2003;112:1011-1012.

[10] Bucchini L, Goldman LR. Starlink corn: a risk analysis. Environmental Health Perspectives 2002;110:5-13.

[11] Bindslev-Jensen C, Ballmer-Weber BK, Bengtsson U, Blanco C, Ebner C, Huorrihane J, Knulst AC, Moneret-Vautrin DA, et al. Standardization of food challenges in patients with immediate reactions to foods – position paper from the European Academy of Allergology and Clinical Immunology. Allergy 2004;59:690-697.

[12] Sampson HA. Utility of food-specific IgE concentrations in predicting symptomatic food allergy. J Allergy Clin Immunol 2001;107:891-896.

[13] Hill DJ, Heine RG, Hosking CS. The diagnostic value of skin prick testing in children with food allergy. Pediatr Allergy Immunol 2004;15:435-441.

Index

A

abnormalities, 8, 22, 62, 73, 83, 102, 167, 168, 172, 173, 175
absorption, 75, 89, 163
accidents, 83, 158
accuracy, 178
achievement, 20
acidity, 40
activated receptors, 101
activation, 38, 94, 100, 101, 115, 133
acute, 3, 4, 5, 12, 27, 30, 57, 94, 95, 100, 102, 107, 108, 111, 114, 159, 160, 164, 174
acute coronary syndrome, 5, 27
acute infection, 95
adenocarcinomas, 31
adhesion, 10, 94, 112, 129
adhesive interaction, 174
adhesives, xii, 177
adipose, 11, 29
adipose tissue, 11, 29
adjustment, 7, 13, 17, 125, 129
administration, 168, 170, 173
adolescence, 137
ADP, 12, 115
adult, 16, 17, 21, 46, 137, 180
adult population, 21
adulthood, 23
adults, xii, 8, 10, 17, 19, 22, 23, 29, 33, 39, 49, 53, 54, 59, 62, 81, 83, 85, 88, 131, 133, 137, 140, 178, 179
aerodigestive tract, 133
affective disorder, 19
African American, 52, 142
African Americans, 52
age, vii, xi, 1, 3, 4, 7, 8, 11, 12, 17, 18, 24, 32, 33, 34, 35, 38, 39, 44, 52, 56, 64, 65, 66, 71, 74, 81, 88, 90, 91, 120, 121, 122, 123, 125, 129, 135, 137, 142, 143, 144, 145, 147, 149, 150, 151, 153, 154, 161, 178
ageing, 50, 130, 133
agent, 22, 162
agents, 82, 159, 173
agricultural, 158
agriculture, 82, 85, 86, 87
AIDS, xi, 157, 159
air, 83, 139, 151
alanine, 170
alanine aminotransferase, 170
Alaska, 41, 57, 158, 160, 164
albumin, 95, 96, 98, 100, 101, 102
alcohol, 41, 129, 130
alcohol consumption, 129, 130
algae, 42, 159, 161
Algal, 48
allele, 15, 20, 52
allergens, xii, 177
allergic reaction, xii, 178, 180
allergy, xii, 171, 178, 179, 180
alpha, ix, 2, 10, 26, 42, 49, 57, 59, 94, 95, 96, 97, 98, 103, 113, 115, 123, 129, 134, 169, 170, 171, 174
alpha-linolenic acid, 2, 26, 42, 49, 59, 113, 123, 134, 171, 174
alternative, 39, 62, 70, 168, 170
alternatives, 39
alters, 103
aluminum, 40
alveolar macrophage, 16
Alzheimer disease, 33, 34, 52, 55, 56, 57, 130, 161

Alzheimer's disease, vii, 1, 2, 19, 20, 39, 40`1, 50, 52, 55, 58
amalgam, 136
American Heart Association, 23, 24, 137, 153
American Indian, 41, 52
American Indians, 41, 52
amines, 14
amino, vii, 37, 38, 45, 46, 48
amino acid, vii, 37, 38, 45, 46, 48
amyloid, 10, 20, 174
amyloid deposits, 20
analysis of variance, 64
analytic techniques, x, 105
anaphylaxis, xii, 178, 180
angina, 29, 106, 107, 114
angiogenesis, 175
angiogenic, 128
angiography, 56
angioplasty, 11
animal diseases, 82
animal models, 159, 171
animal studies, 125, 127, 170
animals, 3, 22, 80, 82, 94, 101, 158, 161, 170
anthropogenic, 136
antiarrhythmic, vii, 1, 6, 80, 112
anti-atherogenic, 106
antibacterial, 51
antibiotics, 82
antibody, 91
antigen, 115, 128, 133
anti-inflammatory eicosanoids, 170
antimony, 120
antioxidant, 29, 83, 111, 174
antitoxin, 159
anxiety, 19
anxiety disorder, 19
APC, 30
APOE, 33, 52, 53
apoptosis, 12
application, 88
aquaculture, ix, xi, 39, 51, 77, 78, 85, 86, 87, 88, 89, 91, 120, 157, 158
arachidonic acid, ix, 13, 14, 16, 17, 38, 81, 94, 96, 101, 169
Arctic, 54
Argentina, 45
arrhythmia, 6, 27
arrhythmias, 2, 114
arsenic, 22, 120
ART, 103, 113
arteries, 6, 45, 80, 123
artery, vii, 1, 3, 5, 6, 9, 11, 25, 26, 29, 80, 91, 106, 108, 115
arthritis, 16
aryl hydrocarbon receptor, 85
ascorbic, 42
ascorbic acid, 42
Asia, 50, 59, 131, 170
Asian, 39, 121, 139, 142, 147, 148, 149, 154
aspirin, xii, 7, 113, 177
assessment, 51, 53, 64, 74, 75, 90, 139
assumptions, 151
asthma, vii, 1, 2, 16, 17, 32, 51, 57, 81, 90, 180
ataxia, 22, 83
Athens, 1
atherogenesis, 94, 100
atherosclerosis, ix, 6, 9, 27, 29, 43, 80, 91, 93, 94, 95, 100, 101, 102, 111, 112, 114, 119
Atherosclerosis Risk in Communities, 32
atherosclerotic plaque, 123
atherosclerotic vascular disease, 100, 101
atherothrombotic, 112
Atlantic, 51, 57
atmosphere, 22
atmospheric deposition, 136
atoms, 79
atopic asthma, 90
atopic dermatitis, 178
atopy, 57
ATPase, 112, 162
attacks, 17
attitudes, 150, 154
Australia, 45, 78, 81
Austria, 54
authority, 87
autoimmune, 2, 95
autoimmune disorders, 95
availability, vii, 37, 40, 46, 70, 86, 122
avoidance, 24, 88, 138
awareness, viii, xi, 61, 66, 67, 70, 71, 72, 73, 76, 135, 136, 138, 143, 144, 145, 150, 154
axonal, 17

B

babies, 162
bacteria, xi, 82, 157, 159
balance sheet, 89
Bangladesh, 88
barley, xii, 177

barriers, 22, 49, 83
batteries, 82, 120
beaches, 64, 83, 160
beef, 85, 91
beer, xii, 177
behavior, 19, 64
behavioral change, 150
beliefs, 54, 154
beneficial effect, vii, 1, 2, 3, 4, 5, 6, 8, 9, 17, 32, 41, 43, 111, 112, 120, 123, 131
benefits, vii, viii, x, xi, 1, 2, 5, 8, 10, 11, 18, 23, 24, 25, 27, 30, 34, 35, 37, 38, 39, 43, 44, 46, 51, 52, 56, 61, 62, 63, 64, 66, 67, 69, 70, 71, 72, 73, 75, 76, 77, 78, 80, 81, 86, 87, 88, 90, 106, 110, 112, 113, 123, 127, 129, 135, 136, 137, 138, 150, 153, 163, 167, 172
beta-carotene, 55
bias, 72, 129, 152
bile, 159, 169, 170
bilge, 158
biliary cirrhosis, 168
bilirubin, 171
binding, 12, 57, 80, 83, 112, 115, 121, 131, 170, 179
bioaccumulation, 121, 163
bioassay, 160, 162, 164
bioavailability, 75, 89
biodegradable, 82
biomarker, 124, 128, 152, 159
biomarkers, x, 5, 43, 103, 118, 123, 124, 128, 137, 139, 159
biomass, 38
biomonitoring, 25, 131, 153
biosynthesis, 14, 90
biota, 136
biotic, 22
bipolar, 19
bipolar disorder, 19
birds, 82
birth, 18, 21, 34, 152, 168
birth weight, 18, 152, 168
birthweight, 137
bivalve, 159, 165
black women, 8
bladder, 41, 53, 124
bladder cancer, 53
bleaching, 23, 84
bleeding, xi, 9, 123, 167
bleeding time, 9
blindness, 18, 81, 172
blood, 4, 6, 8, 9, 10, 12, 17, 20, 22, 28, 29, 32, 33, 35, 38, 39, 54, 57, 80, 95, 96, 98, 99, 101, 103, 114, 115, 121, 123, 131, 136, 137, 138, 139, 140, 141, 142, 150, 151, 152, 153, 154, 168, 173
blood glucose, 8, 28
blood pressure, 9, 28, 29, 57, 123, 136
blood pressure reduction, 9
blood stream, 168
blood vessels, 8, 20
bloodstream, 171
BMI, ix, 94, 95, 96, 99, 100, 129, 147, 150
body fat, 159, 162
body mass index, ix, 94, 99, 129, 149
body weight, 137, 138, 139
boiling, 15, 120
bonds, 79
borderline, 13, 19
borderline personality disorder, 19
Boston, 44, 59, 167, 171
bowel, 160, 167, 168, 173, 175
boys, 178
brain, vii, 1, 2, 6, 17, 20, 22, 37, 38, 42, 50
brain damage, 18
brain development, vii, 1, 2, 17, 20, 42, 50
brain size, 38
Brazil, 45, 87
breakdown, 84, 94, 100
breast cancer, 13, 14, 25, 30, 31, 41, 46, 50, 51, 58, 125, 126, 131, 132
breast carcinoma, 52
breast milk, 22, 81
breastfeeding, 143
brevis, 161
Britain, 121
British Columbia, 54, 121, 131
brominated flame retardants, 22
bronchitis, 16
bronchoconstriction, 90
bronchodilator, 17
bronchospasm, 161
Brussels, 47
burning, 120, 158
burns, 160
bypass, 11, 108
by-products, 84, 121

C

CAD, 106, 108, 109
cadmium, 22, 82, 120, 121, 131, 158

calcification, 45, 100
calcium, viii, 2, 40, 43, 44, 50, 52, 54, 56, 57, 77, 81, 158
caloric intake, 168
calorie, 158
campaigns, 63
Canada, 47, 49, 57, 121, 126, 131, 132, 161, 177
cancer, vii, ix, x, xi, 1, 2, 12, 13, 14, 15, 25, 30, 31, 34, 39, 41, 43, 51, 52, 53, 54, 56, 58, 77, 81, 117, 118, 119, 122, 123, 124, 125, 126, 127, 128, 129, 130, 131, 132, 133, 157, 160, 164, 174
CAP, 178
carbon, 79
carbon atoms, 79
carcinogen, 23, 84
carcinogenesis, 12, 30, 51, 119, 126
carcinogenic, 15, 23, 24, 39, 85, 127
carcinogenicity, 84, 120
carcinogens, 85, 120
carcinoma, 15, 31, 52, 59, 119, 130
carcinomas, 127
cardiac arrest, 5, 6, 27, 103, 160
cardiac arrhythmia, 2, 6
cardiomyopathy, 55
cardiovascular disease, vii, ix, x, 1, 2, 4, 5, 6, 8, 10, 12, 23, 26, 28, 29, 30, 34, 47, 58, 90, 94, 96, 100, 101, 103, 105, 106, 107, 111, 113, 119, 133
cardiovascular risk, ix, 8, 11, 12, 42, 49, 94, 101, 102, 115, 151
cardiovascular system, 22, 115
carotene, 42
carotid arteries, 95
catalase, 12
catfish, 11, 24, 91, 122
cathepsin B, 102
cation, 40
causality, 11, 19, 52
cell, viii, 9, 12, 15, 27, 77, 80, 89, 100, 101, 103, 111, 112, 121, 127, 129, 169, 172
cell adhesion, 10, 112, 129
cell growth, 129
cell line, 127
cell lines, 127
cell membranes, 12, 80, 101, 111, 169
Centers for Disease Control (CDC), 123, 151
central nervous system, vii, 1, 17, 19, 33
Cerastoderma edule, xi, 157, 164
cereals, xii, 40, 46, 177, 180
cerebral arteries, 6
cerebral palsy, 18, 84
cerebrospinal fluid, 20
channels, 83, 112, 115
chemicals, viii, 21, 69, 70, 77, 81, 82, 84, 86, 87, 120
chewing, xii, 177
chicken, 30, 85
chickens, 42
CHILD, 153
child development, 18, 33, 54, 152
childbearing, vii, 1, 24, 32, 34, 35, 38, 39, 88, 151, 153
childhood, 16, 18, 32, 48, 57, 59, 84, 85, 119, 136, 137, 172
children, vii, viii, xii, 1, 16, 18, 22, 23, 24, 54, 56, 62, 64, 71, 72, 77, 84, 85, 88, 123, 125, 137, 140, 145, 151, 152, 160, 168, 169, 171, 173, 175, 178, 179, 180
Chile, 161, 165
China, xi, 4, 26, 46, 49, 121, 130, 133, 157, 158
chlamydia, 102
chlordane, viii, 77, 81, 84, 138
chloride, 84
chlorination, 120
chlorine, 82
chlorophenol, 82
chocolate, 83
cholecalciferol, vii, 37, 44, 53, 56
cholelithiasis, 168
cholestasis, 168, 169, 170, 171, 173, 174
cholestatic liver disease, 173
cholesterol, 2, 8, 10, 62, 69, 95, 96, 97, 98, 114, 169
chromatography, 95
chromium, 82, 120
chronic disease, 30, 40, 45, 48, 57, 58, 91
chronic illness, 82
chronic obstructive pulmonary disease, vii, 1, 32
chronic renal failure, 94, 102, 103
cigarette smoking, 15, 16, 31
ciguatera, xi, 157, 159
citizens, 118
citrus, 82
classes, xii, 112, 178
classical, 100
classification, 109, 110, 146
cleaning, 82
climate change, 159
clinical assessment, 133
clinical presentation, 160
clinical trial, ix, 25, 49, 80, 105, 106, 114, 137, 152, 172

coagulation, 12
coagulopathy, 171
coastal areas, 39, 63, 64, 122
Cochrane, ix, 105, 107, 113
Cochrane Database of Systematic Reviews, ix, 105, 107, 113
cognition, 20, 33, 34, 136, 151
cognitive deficit, 18
cognitive development, 17, 18, 32, 62, 74, 75, 81, 150, 151
cognitive function, 17, 19, 20, 137
cognitive performance, vii, 1, 19, 33, 59
cohort, x, xi, 3, 5, 7, 8, 10, 13, 14, 15, 18, 26, 28, 29, 30, 31, 33, 34, 48, 49, 51, 54, 56, 110, 111, 113, 117, 118, 121, 122, 123, 126, 127, 130, 131, 132, 135, 137, 143, 147, 151, 154, 171
collaboration, 87, 107
college students, 74, 145, 154
colon, x, 12, 13, 30, 41, 43, 51, 117, 118, 126, 132, 159, 162
colon cancer, 13, 30, 43, 51, 126, 132, 162
colon polyps, 159
colorectal cancer, xi, 13, 30, 43, 50, 51, 52, 55, 56, 58, 59, 125, 126, 131, 132, 157, 164
coma, 161
combined effect, 152
commodity, 78
commodity markets, 78
communication, 64, 70, 73, 74, 154
communities, 56, 158
community, 28, 33, 56, 90
comorbidity, 100
competition, 82, 86
complementarity, 45
compliance, xi, 74, 106, 135, 143, 146, 148, 149
complications, xi, 89, 123, 167, 168, 169, 172, 173
components, vii, xii, 1, 2, 13, 40, 41, 79, 96, 98, 125, 169, 177
composition, xii, 26, 53, 88, 89, 95, 103, 106, 130, 168, 169, 172, 174, 177
compounds, vii, 1, 12, 21, 22, 23, 82, 83, 91, 111, 115, 159
comprehension, 17
concealment, 107
concentration, 22, 29, 56, 94, 96, 100, 102, 137, 138, 142, 152, 160, 170
confidence, 38, 43, 107, 109, 110, 125, 126, 138, 142, 146, 148
confidence interval, 38, 43, 107, 109, 110, 125, 126, 138, 142, 146, 148
confidence intervals, 107, 142, 146, 148
conflict, 134
conflict of interest, 134
confounding variables, 129
confusion, x, 24, 88, 135, 138
congestive heart failure, 3, 100
Congress, 91, 151
Connecticut, 53
consciousness, 87
consent, 95, 146
conservation, 86
construction, 84
consumer choice, 62, 138
consumers, 4, 6, 10, 19, 21, 24, 34, 62, 70, 71, 75, 87, 119, 122, 129, 138, 147
consumption patterns, vii, xii, 61, 70, 177
contaminants, viii, 17, 22, 23, 25, 34, 35, 53, 56, 61, 62, 63, 64, 66, 67, 68, 69, 70, 71, 72, 73, 81, 83, 85, 88, 89, 90, 91, 111, 120, 138
contaminated food, 160
contamination, x, 11, 18, 83, 87, 88, 135, 138, 164
control, x, 6, 9, 11, 13, 14, 16, 27, 28, 29, 31, 41, 49, 50, 56, 57, 58, 82, 84, 86, 87, 89, 95, 96, 98, 109, 111, 112, 117, 118, 123, 125, 126, 127, 130, 132, 133, 161, 162
control group, 98, 111
controlled studies, 43
controlled trials, 49, 106, 113, 114, 137
conversion, 42
cooking, 14, 15, 22, 23, 62, 85, 120, 122, 123, 127, 179
cooling, 82
COPD, 57
copper, 40, 82
corn, xii, 43, 84, 114, 177, 178, 179, 180
coronary angioplasty, 108, 114
coronary artery bypass graft, 11, 108
coronary artery disease, vii, 1, 3, 5, 6, 9, 25, 26, 29, 91, 108, 115
coronary heart disease, 2, 3, 5, 10, 12, 25, 26, 27, 29, 30, 32, 42, 46, 52, 59, 75, 80, 88, 103, 113, 115
correlation, ix, xi, 20, 94, 95, 96, 101, 121, 125, 126, 141, 157
correlation analysis, 95
correlation coefficient, 141
correlations, 40, 41, 95
cosmetics, xii, 136, 177
costs, 158
cough, 162
covering, 126

COX-2, 14, 31, 127, 132
crab, 24
C-reactive protein, ix, 10, 94, 95, 102, 128, 133, 174
crops, 84
cross-sectional, 21, 56, 81, 101
cross-sectional study, 56, 81, 101
CRP, ix, 94, 95, 96, 98, 100, 102, 128, 129, 133
crude oil, 83
crustaceans, 85, 86, 121
culture, 87, 158, 163
CVD, 43, 94, 97, 100
cyanobacteria, xi, 158
cycles, 11, 22
cyclooxygenase, 17, 101, 174
cyclooxygenase-2, 174
cytokine, 94, 102, 115, 128, 129, 174
cytokines, 10, 94, 100, 101, 103, 170
cytotoxic, 159, 162, 174
cytotoxic action, 162

D

dairy, 55, 85, 126, 130, 180
dairy products, 55, 85, 126, 130, 180
data collection, 95
database, 107, 123, 124, 153
DBP, 99
de novo, 170
deafness, 18, 84
death, viii, 2, 3, 6, 7, 10, 11, 12, 26, 27, 29, 30, 38, 56, 77, 80, 81, 84, 88, 90, 94, 100, 103, 106, 111, 112, 113, 114, 115, 118, 123, 160, 161
death rate, 7, 160
deaths, ix, x, 11, 105, 106, 109, 113, 119, 161
decisions, vii, 61, 63, 64, 71, 72, 73
defibrillator, 108
deficiency, xi, 2, 53, 58, 59, 95, 101, 104, 167, 168, 173, 175
deficit, 19, 78, 152
deficits, 2, 18, 22
definition, 81, 85
degrading, 87
delivery, 9, 140, 145, 146, 150
dementia, 2, 19, 20, 33, 48, 52, 53, 57, 59, 81
demographics, 43
Denmark, 10, 118, 122
density, 55
Department of Agriculture, 45, 58
Department of Health and Human Services, 88, 152, 155
deposition, 120, 136
deposits, 20, 80
depressed, 20
depression, vii, 1, 19, 20, 21, 34, 81, 91
depressive disorder, 19
depressive symptoms, 21, 34
deprivation, 52
derivatives, 159, 160
dermatologic, 172
developed countries, 12, 78
developing brain, 17, 43, 56, 153
developing countries, 78, 86
developmental deficits, viii, 77, 85
dextrose, 170
diabetes, vii, viii, ix, 1, 2, 8, 23, 29, 46, 47, 48, 54, 56, 57, 59, 77, 80, 89, 93, 95
diabetes mellitus, ix, 47, 48, 93
diabetic patients, 9
diabetic retinopathy, 10
dialysis, ix, 93, 94, 95, 99, 100, 102
Diamond, 175
diarrhea, xi, xii, 157, 178
diarrhoea, 159, 162
diastolic blood pressure, 8, 9, 96, 98, 99, 101
diastolic pressure, 95
dibenzofurans, 84, 91
dibenzo-p-dioxins, 23, 34
diesel, 83
diet, vii, viii, xii, 1, 2, 9, 10, 15, 16, 18, 19, 20, 21, 25, 26, 29, 33, 34, 40, 41, 42, 43, 44, 45, 51, 57, 58, 77, 80, 89, 91, 103, 106, 110, 111, 112, 113, 115, 118, 121, 123, 126, 127, 128, 130, 131, 132, 133, 138, 143, 170, 174, 177, 179
dietary, ix, x, 2, 5, 9, 11, 13, 14, 16, 17, 20, 21, 23, 24, 27, 29, 30, 31, 32, 33, 34, 40, 42, 45, 46, 51, 52, 54, 55, 56, 76, 80, 84, 85, 90, 91, 94, 103, 105, 106, 109, 111, 113, 114, 117, 121, 122, 125, 130, 132, 133, 137, 138, 146, 153, 163, 174, 175
dietary fat, 13, 90, 125
dietary habits, 20
dietary intake, 5, 16, 17, 21, 27, 31, 32, 55, 80, 94, 121, 122, 163, 175
dietary supplementation, x, 52, 105, 109, 113, 174
diets, 13, 29, 40, 43, 45, 111, 115, 118
differentiation, 17
digestive tract, 12, 40, 41, 160
dinoflagellates, 160
dioxin, 23, 34, 85, 90, 121, 131
dioxin-like PCBs, 131

dioxins, vii, 1, 22, 23, 24, 34, 39, 85, 91, 111, 120, 121, 131, 138, 158
disease rate, 45
diseases, ix, xi, 4, 19, 45, 46, 47, 82, 93, 119, 157
disorder, 19, 96
disposition, 89
distillation, 83
distribution, 15, 22, 160, 165
diversification, 86
division, 17
docosahexaenoic acid, ix, 2, 5, 10, 17, 19, 21, 30, 33, 37, 38, 48, 49, 50, 55, 56, 57, 80, 101, 105, 106, 107, 108, 127, 130, 169, 170
Doppler, 95
dosage, 107, 109, 178
dose-response relationship, 7
dosing, 53
double bonds, 79
drug therapy, 9
drugs, 57, 113
dumping, 82, 158
duplication, xi, 157
duration, ix, 6, 43, 95, 105, 107, 109, 120, 163, 168
dusts, 82
dysarthria, 18
dyslipidemia, 9, 102
dyspnea, 162

E

earth, 45, 76
Eastern Europe, 42, 59
eating, vii, viii, 2, 3, 5, 6, 7, 14, 16, 20, 21, 23, 24, 34, 37, 38, 39, 41, 42, 46, 61, 64, 69, 70, 71, 72, 75, 77, 81, 110, 111, 113, 130, 137, 159, 160, 161
echinoderms, 165
ecological, 2, 21, 25, 29, 30, 34, 40, 43, 131
ecology, 88
economic growth, 86
ecosystems, 73, 86, 87, 88
education, 65, 66, 148
egg, 45, 54, 168, 179
eicosanoids, 14, 38, 101, 112, 169, 174
eicosapentaenoic acid, ix, 2, 5, 15, 32, 37, 38, 80, 101, 105, 106, 107, 108, 114, 115, 116, 118, 169, 170
elaboration, xii, 177
elderly, 5, 8, 19, 27, 28, 33, 34, 50, 113, 119
elderly population, 19, 27, 113
electrical system, 84
ELISA, 95, 162
Embase, ix, 105, 107
emphysema, 16
employees, 145
empowered, 87
emulsifier, 168
emulsifying agents, 173
emulsions, xi, 167, 168, 169, 170, 171, 172, 173, 174
encephalitis, 82
endocrine, 2, 23, 128, 133
endocrine disorders, 2
endocrine system, 128
endometrial cancer, 14, 31, 58, 124
endosperm, xii, 178
endothelial cell, 12, 112, 115
endothelial dysfunction, ix, 94, 100, 112, 133
endothelium, 42, 94, 101, 115, 174
end-stage renal disease, ix, 93, 94
energy, 2, 9, 38, 45, 46, 81, 94, 167
environment, 62, 69, 73, 81, 83, 84, 86, 87, 88, 91, 121, 163
environmental contaminants, 24
environmental contamination, 83, 88
environmental factors, 52
environmental protection, 71
Environmental Protection Agency (EPA), ix, 2, 4, 10, 19, 32, 37, 38, 43, 50, 72, 80, 83, 84, 91, 96, 101, 105, 106, 108, 109, 112, 122, 123, 137, 138, 139, 149, 151, 152, 153, 155, 169, 170, 171
enzyme-linked immunosorbent assay, 95
enzymes, 83, 101, 174
epidemic, 48
epidemiologic studies, 3, 106, 118
epidemiology, 53, 172
epigenetic, 19
epigenetic mechanism, 19
ergocalciferol, 44, 53
erythrocyte, 18, 19, 95, 96, 98, 103, 174
erythrocyte membranes, 19
esophageal cancer, 133
esophagus, x, 12, 41, 117
essential fatty acids, 19, 21, 101, 169, 170, 172
esters, 10, 95
estrogen, 14, 42, 50
estuaries, viii, xi, 61, 63, 71, 72, 81, 158, 163
Ethiopian, 56
ethnic groups, 70, 121, 139
ethnicity, 64, 74, 139, 142, 148, 149
ethylene, 82
etiologic factor, ix, 93

etiology, 52, 168, 173
Euro, 165
Europe, xi, 13, 42, 122, 157, 161, 162, 168, 170
evolution, 33, 50, 54
exaggeration, 112
examinations, 95
exclusion, 179
excretion, 169
exercise, 90
exposure, viii, 18, 22, 23, 24, 30, 32, 33, 34, 59, 63, 74, 75, 77, 83, 84, 85, 89, 90, 119, 120, 121, 122, 131, 136, 137, 138, 139, 144, 149, 150, 151, 152, 153, 154, 159, 163, 165
Exposure, 75, 119, 152
extraction, 124, 161, 163
Exxon, 158
Exxon Valdez, 158

F

factor analysis, 95, 96
factorial, 168
FAD, 171
failure, 103, 119, 168, 172, 173
family, xii, 63, 71, 143, 177
FAO, 30, 78, 85, 86, 87, 89, 90, 122
farmers, xii, 178
farming, 85, 86, 158, 163
fast food, 11
fasting, 8
fat, xi, 2, 9, 14, 23, 26, 30, 31, 34, 40, 42, 50, 58, 62, 83, 84, 85, 89, 91, 103, 111, 113, 125, 126, 131, 132, 133, 158, 167, 170, 171, 173, 175
fat soluble, 84, 111
fatal arrhythmia, 114
fats, viii, 16, 77, 118
FDA, 63, 70, 71, 72, 75, 84, 85, 91, 122, 137, 143, 144, 145, 146, 149, 151, 164, 171
FDA approval, 171
fear, 168
feces, 22
feeding, 22, 58, 85, 90, 162, 167
females, ix, 21, 93
fetal, xi, 17, 18, 22, 25, 38, 135, 136, 139, 149, 150, 153
fetal brain, 17
fetal growth, 18, 25, 153
fetal growth retardation, 25, 153
fetal tissue, 136
fetus, 17, 42
fetuses, 62, 137
fever, 162
fibrinogen, 9
fibrosis, 168
filtration, xi, 157
finfish, 79
Finland, 3, 21, 26, 30, 34
fires, 158
fish meal, 5, 6, 9, 11, 16, 20, 27, 140, 142
fish oil, ix, xi, 16, 25, 26, 27, 30, 34, 44, 50, 54, 69, 80, 81, 90, 93, 94, 96, 103, 105, 106, 107, 109, 111, 114, 115, 118, 123, 124, 125, 126, 127, 130, 132, 134, 137, 152, 167, 170, 171, 172, 173, 174, 175
fish production, 86, 87
fisheries, 78, 81, 85, 86, 87, 89
fishers, 88
fishing, 62, 63, 64, 70, 71, 74, 76, 87, 121, 144
flavonoids, 11
flood, 83
flow, 124, 170
fluid, 20, 82
fluorine, 82
folliculitis, 171
food, xii, 11, 14, 26, 38, 44, 45, 49, 54, 57, 62, 69, 73, 75, 78, 79, 82, 83, 84, 85, 86, 87, 88, 89, 91, 120, 121, 127, 130, 136, 143, 145, 155, 158, 160, 177, 178, 179, 180
food allergy, 179, 180
Food and Drug Administration (FDA), 35, 63, 75, 122, 137, 145
food production, 85
food products, xii, 177
foodstuffs, 158, 163
Ford, 39, 51, 54
forestry, 86
forests, 40
fortification, 49
fossil, 38, 120
fossil fuel, 38, 120
fractures, 43, 53, 54
France, 161, 164
free radicals, 12, 40, 83
freedom, 163
fresh water, 81, 160
freshwater, xi, 12, 24, 39, 56, 63, 85, 88, 111, 136, 140, 158
fruits, 126
frying, 11, 23, 120
fuel, 38, 83

funding, 47, 74
fungal, 82

G

gallbladder, 41, 46
gallbladder disease, 46
gas, 95
gas chromatograph, 95
gastric, 41, 58, 119, 127
gastrointestinal, 136, 161, 167, 174
gastrointestinal tract, 136, 167
gender, 8, 21, 64, 76, 144, 145
gender differences, 145
gene, 14, 30, 80, 85, 115, 127
gene expression, 85
general knowledge, 66, 71
generation, 40, 86, 174
genes, 38, 91, 112
Geneva, 30, 91
genotoxic, 130
Germany, 45, 108, 122, 168, 171
gestation, 17, 137, 143, 146, 150
Gibbs, 54
girls, 178
global trade, 78
GLP-1, 115
glucose, 8, 9, 28, 29, 112, 115
glutamate, 161
glutathione, 12, 42, 170
glutathione peroxidase, 12
glycemic index, 46
glycolysis, 112
gold, 25, 168, 178
gold standard, 168, 178
governance, 87
government, iv, 87, 123, 136
GPO, 151, 152
grafting, 11, 108
grain, 44, 46
grains, 39, 40, 46, 129
grass, xii, 177
gray matter, 17
Great Lakes, 18, 33, 91
Greece, 1, 42, 58, 154
Greenland, 3, 16, 26, 80, 88, 106, 118, 130
ground water, 82
grouping, 78
groups, x, 9, 70, 83, 86, 95, 97, 106, 109, 121, 123, 128, 139, 142, 144, 146, 147, 150, 154, 170
growth, viii, 18, 22, 25, 43, 56, 77, 78, 80, 81, 86, 103, 112, 125, 128, 129, 133, 153, 170, 171, 172, 174, 175
growth factor, 112, 133
growth rate, 22, 171
Guatemala, 160, 164
guidance, 35, 73, 84, 85, 138
guidelines, xi, 23, 107, 121, 122, 135, 149, 178

H

habitat, 22
Haiti, 159
half-life, 22, 84, 150
halibut, 11, 24, 147
hallucinations, 161
halogenated, 2
handling, xii, 178
hardwood forest, 40
harm, vii, 1, 19, 43, 81, 136
harmful effects, 123
harmony, 43
Hawaii, 40, 59
hazards, 81
HDL, 8, 95, 96, 97, 98, 114
health, iv, vii, viii, xi, 1, 2, 7, 8, 18, 19, 21, 22, 24, 25, 28, 35, 37, 38, 39, 43, 44, 45, 46, 49, 52, 53, 54, 56, 57, 59, 61, 62, 63, 67, 69, 70, 71, 72, 73, 74, 75, 76, 77, 78, 80, 81, 84, 86, 87, 90, 91, 94, 103, 112, 118, 119, 121, 122, 127, 129, 130, 131, 137, 138, 143, 145, 153, 154, 157, 159, 164, 165
Health and Human Services, 152, 155
health care, 7, 143
health care professionals, 7
health effects, viii, 8, 22, 38, 39, 77, 94, 137, 143
health problems, xi, 18, 157, 159
healthcare, 123
heart, viii, 2, 3, 5, 6, 8, 9, 10, 11, 12, 23, 24, 25, 26, 27, 29, 30, 32, 42, 46, 47, 52, 55, 56, 59, 62, 69, 75, 77, 80, 88, 90, 91, 100, 103, 113, 115, 118, 123, 136, 137, 153
heart attack, viii, 5, 77, 123
heart disease, 2, 3, 5, 10, 11, 12, 25, 26, 27, 29, 30, 32, 42, 46, 52, 59, 62, 75, 80, 88, 91, 113, 118, 119, 123
heart failure, 47, 56, 59
heart rate, 6, 9, 136
heat, 81, 84
heat transfer, 84

heavy metal, vii, 1, 11, 21, 22, 34, 62, 82, 83, 87, 88, 121
heavy metals, vii, 1, 11, 21, 34, 62, 82, 87, 88, 121
height, 44, 46
hemisphere, 121
hemodialysis, ix, 93, 94, 95, 102, 103
hemorrhage, 8
hemorrhagic stroke, 6, 8
hemostatic, 112
hepatic fibrosis, 168
hepatotoxins, xi, 158
herbicides, 23, 82
herring, 11, 23, 24, 78
heterogeneity, 80, 106, 107, 109, 114
Higgs, 39, 51, 54
high risk, x, 10, 91, 95, 100, 105, 107, 109, 113
high school, xi, 65, 66, 135, 143
high temperature, 14
higher education, 20
high-risk, 10, 114, 137, 173
hip, 6, 44, 48, 51, 55
hip fracture, 44, 48, 51, 55
hippocampal, 164
hippocampus, 161
Hispanic, 65, 139, 148
Hispanics, 52
histamine, 159
histology, 170
homocysteine, 54
Hong Kong, 119, 130
hormone, 31, 53, 55, 112, 129, 132
hormones, 112
hospital, 160
hospitalization, 160, 162
hospitalizations, 160
hostility, 19
household, 64
household income, 64
HPLC, 161, 162, 164
human, xii, 2, 20, 21, 22, 23, 25, 45, 49, 53, 54, 55, 56, 59, 80, 81, 82, 84, 85, 86, 87, 90, 112, 114, 115, 120, 121, 127, 130, 131, 132, 136, 150, 159, 162, 164, 165, 171, 173, 175, 177
human exposure, 84, 120, 121, 130, 131
humans, xi, 10, 22, 38, 42, 62, 73, 78, 85, 94, 101, 121, 125, 136, 137, 157, 158, 161, 170, 174
Hungary, 45
hydraulic fluids, 84
hydro, 14, 82, 83, 158
hydrocarbon, 15, 82, 83, 85, 120
hydrocortisone, 115
hydrogen, 82
hyperactivity, 19, 151
hypercalcemia, 53
hypercholesterolemia, 108, 109
hyperhomocysteinemia, 45
hyperlipidemia, 113, 116
hypertension, 2, 9, 29, 47, 51, 57
hypertensive, 9, 29
hypertriglyceridemia, xi, 8, 167, 171
hypotension, 160
hypotensive, vii, 1
hypothalamic, 10
hypothesis, 3, 16, 40, 55, 127, 150, 160, 170

I

IAEA, 89
IARC, 130
ICAM, 10, 112
ICD, 108
dentification, xii, 47, 130, 177
identity, 64
IgE, xii, 178, 179, 180
IGF, 129
IGF-I, 129
IL-1, 10, 112
IL-2, 10
IL-6, 94, 95, 96, 97, 98, 100, 101, 112, 128, 129
IL-8, 112
immune function, 129
immune response, 123, 173
immune system, viii, 77, 128
Immunoglobulin E, vi, 177
immunological, 84
immunosuppressive, 125, 170
impairments, 18, 23
implementation, 165
in utero, viii, 18, 47, 50, 77, 83, 84, 137
in vitro, 125, 127, 170, 174
in vivo, 170
inactive, 12
incidence, x, xi, 2, 3, 5, 6, 7, 8, 11, 14, 15, 16, 17, 26, 28, 29, 31, 51, 59, 62, 94, 106, 118, 123, 125, 127, 157, 180
incineration, 23, 84, 158
inclusion, 107, 123, 162
income, 39, 64, 65, 66, 76, 78, 86, 87, 147, 148, 154
incomes, 65
incongruity, 138

India, 45, 46, 57, 108
Indian, 52, 91, 114, 152
indicators, 70, 74, 126, 128
indices, ix, 94, 100, 121, 131, 137
indigenous, 54, 76
indigenous peoples, 54
induction, 55
industrial, 23, 38, 83, 84, 120
industrial accidents, 83
industrial chemicals, 84
industrialized countries, 125
industry, xi, 87, 157
infancy, 17, 18, 136
infant formulas, 81
infants, x, 17, 18, 32, 63, 71, 81, 135, 136, 140, 141, 150, 152, 171, 172, 175
Infants, 145
infarction, 4, 5, 7, 11, 12, 107
infection, 57
infections, 51, 82, 95, 164, 168, 171
infectious, xi, 157, 164
infestations, 39
inflammation, ix, 9, 13, 20, 38, 43, 49, 50, 90, 93, 94, 95, 100, 101, 102, 112, 118, 124, 125, 129, 132, 133, 169, 170, 173
inflammatory, vii, ix, x, 1, 2, 10, 12, 16, 20, 29, 38, 49, 58, 81, 93, 94, 96, 100, 101, 102, 103, 106, 112, 118, 124, 128, 129, 133, 169, 170
inflammatory bowel disease, ix, 93
inflammatory disease, 49, 102
inflammatory mediators, 112, 169
inflammatory response, 10, 12, 100, 112
influenza, 43, 47, 48, 49, 50, 53
influenza a, 43, 48, 49
Information System, 91
informed consent, 95, 146
ingest, 121
ingestion, 22, 75, 83, 122
inhibition, 12, 13, 14, 16, 17, 43, 83, 128, 162, 170, 174
inhibitor, 113
inhibitors, 116
inhibitory, 125
inhibitory effect, 125
initiation, 95, 126
injections, 169
injury, iv, 94, 112
inorganic, 22, 83, 131, 136, 153, 154
insecticides, 22, 120
insects, 82
insight, 173
institutional change, 86
institutions, 146
insulin, ix, 8, 9, 29, 48, 93, 102, 103, 112, 115, 133
insulin resistance, ix, 8, 93, 102, 103, 112
insulin sensitivity, 9, 133
insulin-like growth factor, 133
insulin-like growth factor I, 133
intellectual development, viii, 77, 81
intelligence, 137
intelligence quotient, 137
interaction, 22, 65, 121
interactions, 102, 112
intercellular adhesion molecule, 10
interleukin, ix, 10, 94, 95, 102, 115, 169, 174
interleukin-1, 10, 174
interleukin-6, 10, 95, 102, 169, 174
internet, 67, 68
interval, 38
intervention, 75, 85, 106, 109, 128
interview, 64, 145
interviews, 143, 146
intima, 9
intoxication, 159
intravenous, xi, 167, 168, 169, 171, 172
intravenously, 169
investigations, 164
iodine, 2, 158
ion channels, 83
ionic, 82, 136
ions, 40, 130
IQ, 38, 151
Iraq, 137
Ireland, 123, 161, 165
iron, viii, 2, 40, 58, 77, 81
irritation, 161
ischaemic heart disease, 55
ischemic, 3, 4, 6, 7, 11, 25, 80, 118, 119
ischemic heart disease, 3, 4, 11, 25, 80, 118, 119
ischemic stroke, 6, 7
island, 3, 160
Italy, 5, 26, 41, 90, 108, 165

J

JAMA, 49, 51, 52, 56, 59, 103, 115, 124
Japan, xi, 13, 27, 31, 34, 45, 46, 56, 58, 59, 108, 123, 125, 127, 131, 132, 133, 137, 157, 159, 163
Japanese, 3, 11, 13, 16, 17, 19, 27, 29, 30, 31, 32, 35, 40, 59, 91, 119, 122, 127, 130, 132, 133

Japanese women, 13, 32
Jefferson, 105
joints, 81
Jun, 151, 153, 154, 175
jurisdiction, 87

K

K^+, 162
kappa, 174
kappa B, 174
kerosene, 83
kidney, 2, 41, 100, 123, 131
kidneys, 22, 159
King, 27, 102, 103, 122
Kiribati, 122
Korea, 133
Korean, 141

L

laboratory method, 95
laboratory studies, 43
lactating, 42, 152, 154
lactation, 53, 151
lakes, 22, 75, 82, 83, 122
land, 86
language, x, 17, 18, 107, 117, 136, 138
language development, 18
larynx, 41
Latino, 144, 154
laxatives, xii, 177
LDL, 8, 95, 96, 97, 98
leakage, 83
lean body mass, 100
learning, 23, 33, 40, 143, 145
left ventricular, 6
legislation, xi, 157
legumes, 126, 129
leptin, 9, 29
lesions, 9, 159
leukaemia, 41, 51
leukemia, 122
leukocyte, 9, 16
leukocytes, 94
licenses, 71
licensing, 144
life expectancy, 45
lifespan, 83
lifestyle, 11, 20, 34, 118, 132, 154
lifestyles, 44
life-threatening, 173
lifetime, 2, 137
likelihood, 6, 51, 126
limitations, 45, 179
linear, 21, 41
linear regression, 41
links, 40, 43, 52
linoleic acid, 169, 171
linolenic acid, 43, 103, 106, 114, 168, 173, 175
lipemia, 9
lipid, ix, xi, xii, 11, 12, 28, 30, 53, 89, 93, 101, 112, 113, 114, 115, 116, 159, 167, 168, 169, 170, 171, 172, 173, 174, 175, 177, 180
lipid metabolism, 28
lipid oxidation, 11
lipid peroxidation, ix, 12, 30, 93, 114
lipid profile, 28, 113, 116
lipids, 29, 49, 51, 79, 89, 95, 100, 112, 113, 114, 116, 130, 154, 162, 168, 174, 175
lipooxygenase, 101
lipophilic, 23, 84
lipopolysaccharide, 115, 174
lipoprotein, 8, 43, 102, 106, 112, 113, 174
lipoproteins, 12, 111, 113, 116, 169
lipoxygenase, 17
liquid chromatography, 165
liquidity, 101
liver, viii, xi, 9, 41, 43, 44, 57, 77, 84, 119, 127, 131, 167, 168, 169, 170, 171, 172, 173, 174, 175
liver cancer, 84
liver disease, viii, xi, 77, 167, 168, 172, 173, 174, 175
liver failure, 168
liver function tests, 171
liver transplant, 171
liver transplantation, 171
livestock, 158
living standard, 86
living standards, 86
lobsters, 78
local government, 136
location, 2, 68, 69, 139, 143, 144, 145
London, 57, 154
long period, 159
longevity, 22
longitudinal studies, x, 118
low risk, x, 10, 105, 106, 107, 109
low-density, 43, 102

low-density lipoprotein, 43, 102
lower respiratory tract infection, 57
low-income, 76, 78, 154
low-level, 38, 136, 137
LPS, 174
LTP, xii, 177
lubricants, 83, 84
luminal, 159
lung, x, 15, 16, 31, 41, 50, 58, 91, 117, 120, 124, 127, 133
lung cancer, x, 15, 31, 41, 50, 58, 91, 117, 120, 124, 127, 133
lung disease, 16
lung function, 16
lungs, 16
lupus, 81
lymphocyte, 174
lymphoma, 41, 45, 51, 53, 124
lymphomas, 41
lysine, 45

M

mackerel, 24, 78, 138, 146, 147
macrophage, 174
macrophages, 53
magnesium, 40, 52
Maine, 57
maintenance, 103
maize, xii, 177, 178, 179, 180
major depression, 20, 34
major depressive disorder, 19
malaria, 82
males, ix, 93, 126
malignancy, 95, 126
malnutrition, 94, 95, 96, 100, 101, 102, 103
management, 64, 71, 74, 86, 87, 88, 172, 173, 175
manganese, 40
man-made, viii, 77, 81
mantle, 161
marine environment, 81
marine mammals, 137, 161
market, 38, 54, 78, 165, 168
markets, 160, 164
mass spectrometry, 165
Massachusetts, 139, 143, 145, 164, 167
maternal, xi, 17, 18, 21, 22, 50, 135, 136, 137, 139, 141, 146, 147, 148, 149, 150, 154
maternal age, xi, 135
matrix, 99
meals, 6, 14, 16, 24, 45, 71, 125, 139, 140, 141, 142
meanings, 76
measures, 137
meat, xi, 2, 13, 22, 23, 30, 31, 33, 48, 54, 62, 69, 70, 81, 85, 119, 125, 126, 129, 130, 139, 143, 144, 157, 159, 160, 161
mechanical properties, 101
media, 9, 70, 71, 72, 73, 143
media messages, 72
median, 138, 141, 178
mediators, 112, 169
medication, 81
medicine, 50
Mediterranean, 27, 28, 34, 42, 51, 89, 91, 112, 113, 115, 154
Medline, 138
MEDLINE, ix, 105, 107
melanoma, 41, 51
membranes, 17, 19, 112, 172
memory, 17, 18, 23, 33, 136, 161
memory deficits, 136
men, 3, 6, 7, 8, 9, 10, 12, 14, 15, 19, 21, 26, 28, 29, 30, 33, 40, 42, 48, 51, 53, 59, 80, 111, 114, 115, 119, 127, 128, 129, 133, 143
mental disorder, 29
mental health, 32
mental retardation, 18, 84
mercury, vii, viii, xi, 1, 2, 12, 18, 19, 22, 23, 24, 28, 30, 32, 33, 34, 35, 38, 39, 40, 54, 61, 62, 63, 67, 70, 72, 73, 74, 75, 77, 81, 82, 83, 84, 89, 91, 111, 114, 115, 120, 121, 122, 123, 131, 135, 136, 137, 138, 139, 140, 141, 142, 143, 145, 146, 147, 149, 150, 151, 152, 153, 154, 158, 159, 164
Mercury, 12, 22, 30, 38, 39, 53, 55, 68, 74, 75, 83, 91, 111, 120, 130, 131, 135, 136, 151, 152, 154, 155
messages, 72, 173
meta analysis, 52
meta-analysis, 3, 7, 8, 13, 26, 28, 30, 43, 46, 48, 49, 51, 57, 59, 80, 106, 107, 108, 109, 111, 112, 113, 114, 132
metabolic, xii, 29, 52, 96, 102, 168, 173, 178
metabolic syndrome, 29, 52
metabolism, 9, 10, 16, 21, 28, 29, 32, 38, 94, 101, 112, 130, 174, 175
metabolite, 16, 20
metabolites, 16, 17, 54
metal ions, 40
metals, 21, 69, 82, 123, 131, 158, 159, 163
metastasis, 128

metastatic, 14, 127
metastatic cancer, 14, 127
methyl group, 79
methylation, 121
methylmercury, vii, 1, 11, 17, 18, 19, 21, 22, 23, 24, 30, 32, 33, 34, 38, 63, 76, 83, 84, 121, 131, 136, 137, 138, 139, 149, 151, 152, 153, 154, 159, 164
Mexican, 139
Mexico, xii, 41, 177, 178
Miami, 40
mice, 84, 102, 115, 170, 175
microalgae, 50
microbial, 22, 83
micrograms, 161
micronutrients, 152
microorganisms, 159, 162
micro-organisms, 121
microvascular, 10, 28
microvasculature, 10
middle-aged, 4, 5, 9, 10, 19, 29, 95, 103
migration, 12, 17
milk, 21, 22, 34, 44, 48, 81, 84, 129, 179
mineral oils, 83
minerals, vii, viii, 1, 2, 77, 81, 83
mites, 82
modality, 167
models, 33, 43, 46, 146, 159, 171
modulation, 10, 112, 173
molecules, 10, 55, 112
mollusks, 136, 158
monocyte, 112, 174
monocytes, 115
mononuclear cell, 115
mononuclear cells, 115
monotherapy, xi, 167, 168, 170, 171, 172
monounsaturated fat, 42, 96
monounsaturated fatty acids, 96
mood, vii, 1, 20, 33, 34, 59
mood change, 33
mood disorder, vii, 1
morbidity, 27, 102, 118
morning, 81
morning stiffness, 81
mortality, x, 2, 3, 5, 6, 7, 10, 12, 15, 25, 26, 27, 28, 29, 30, 31, 38, 41, 43, 51, 52, 55, 59, 75, 80, 86, 100, 102, 105, 109, 110, 113, 118, 122, 125, 126, 131, 132, 164, 168, 171
mortality rate, 3, 41, 43, 52, 80, 126, 132
mortality risk, 102
mothers, 16, 21, 24, 38, 138, 143
motor skills, 18
mouse, 83, 160, 162, 164, 173
mouse model, 173
mouth, viii, 61, 73, 160
MUFA, 97, 98, 115
multiple myeloma, 41, 51, 53
multiple sclerosis, 56
multivariate, 43, 95, 98
murine model, 170, 175
muscle, 22, 94, 95, 96, 100, 102, 103, 111, 158
muscle atrophy, 102
muscle cells, 111
mushrooms, 44
mustard oil, 114
mutagen, 159
mutations, 30
Mycobacterium, 55
myeloma, 41, 51, 53
myocardial infarction, 2, 3, 4, 5, 8, 9, 11, 12, 26, 27, 29, 30, 80, 89, 90, 103, 106, 107, 108, 111, 112, 113, 114, 130
myocardium, 3

N

Na^+, 115, 162
NASA, 40
nasopharyngeal carcinoma, 119, 130
National Research Council, 76, 89, 151
Native American, 63, 70, 75, 139, 142, 154
Native Americans, 63
natural, viii, 16, 22, 44, 64, 77, 81, 82, 86, 134, 136, 163
necrosis, ix, 10, 94, 95, 115, 169, 170, 174
negative relation, 6
neonatal, 2, 51, 172
neonates, 63, 168, 172, 173
nerve, 20
nerve cells, 20
nervous system, 2, 17, 20
Netherlands, 26, 30
neural tissue, 172
neurobehavioral, viii, 77
neurodegenerative, 161
neurogenesis, 164
neurological disorder, 32
neuronal cells, 17
neuroprotective, 10
neuropsychiatric disorders, 2, 19, 81, 91
neurotoxic, 22, 161

neurotoxicity, 23, 159, 164
neurotransmitter, 10, 21
neutrophils, 16
New Jersey, v, viii, 52, 61, 63, 64, 65, 66, 73, 74, 138, 145, 153, 154
New Mexico, 41
New York, 54, 61, 63, 64, 65, 66, 71, 72, 73, 76, 91, 121, 131, 135, 145
New Zealand, 78, 137
niacin, viii, 77, 81
nickel, 120
Nielsen, 113
Nigeria, 77
nitrate, 133
nitric oxide, 115
nitrogen, 172
nitrosamines, 120, 130
nitroso compounds, 120, 127
N-N, 120
non-Hodgkin lymphoma, 41, 51
non-invasive, 64
nonsmokers, 147
Norfolk, 28
normal, 101, 143, 171
normalization, 171
norms, 16
North America, 48
North Carolina, 76, 164
Northeast, 39
Norway, 57, 108, 123, 125, 132
NRC, 62, 76, 83, 89
NS, 56, 65, 66, 68, 97, 98, 140, 141
nuclear, 112, 115
nulliparity, xi, 135, 150
nulliparous, 148
nursing, 24, 38, 42, 44, 52, 123, 138
nutrient, 45, 163
nutrients, 21, 54, 136, 167
nutrition, xi, 29, 30, 49, 50, 56, 58, 80, 123, 132, 133, 167, 172, 173, 174, 175

O

obese, 9, 175
obesity, 8, 29, 132, 143, 147
observational design, 19
observations, 19, 95, 159
obstructive lung disease, 16
occlusion, 174
occupational, 83, 131
Oceania, 78
oceans, 22, 87
odds ratio, 4, 16, 107, 109, 110, 111, 126, 146, 148
Ohio, 40
oil, vii, ix, xi, 1, 16, 25, 26, 32, 34, 37, 42, 43, 44, 46, 48, 50, 54, 57, 80, 81, 82, 83, 89, 90, 93, 94, 96, 103, 105, 106, 107, 108, 109, 111, 112, 113, 114, 115, 118, 123, 124, 125, 126, 127, 132, 134, 137, 152, 158, 167, 168, 170, 171, 172, 173, 174, 175
oil spill, 83
oils, viii, xi, 11, 42, 59, 62, 69, 73, 83, 111, 118, 167, 168, 169
older adults, 33, 53, 59, 81, 133
olive, 108, 109, 112, 115
olive oil, 108, 109, 112, 115
omega-6, 42, 56, 58, 112, 169, 171, 174
oncology, 117
online, 153, 165
operator, 178
optimal health, 52, 130
oral, x, 12, 41, 53, 76, 117, 178, 179
oral cavity, x, 12, 41, 117
orange juice, 44
organ, 168
organic, 2, 18, 22, 23, 38, 39, 53, 82, 83, 84, 87, 90, 120, 140, 153, 158
organic compounds, 82, 87, 158
organic matter, 83, 84
organism, 82
organochlorine compounds, 84, 88
organotin compounds, 22
orientation, 86
osteomalacia, 47, 50
osteopenia, 54
osteoporosis, 47, 50
outcome relationship, 18
outpatient, 126
overload, 100
overweight, xi, 28, 29, 103, 135, 147
ownership, 85
oxidation, 11, 43
oxidative, ix, 38, 40, 93, 95, 100, 102, 112
oxidative stress, ix, 38, 40, 93, 95, 100, 102
oxide, 112, 115
oxides, 40
oxygen, 16
oysters, 24, 78
ozone, 40

P

Pacific, 139, 142, 154
Pacific Islander, 139, 142, 154
pain, 178
paints, 120
Pakistan, 45, 46
pancreas, x, 12, 41, 117
pancreatic, 41, 54, 124
pancreatic cancer, 54
paralysis, 160
parameter, 95, 96
parasite, 159
parasites, xi, 54, 157
parathyroid, 53, 55
parathyroid hormone, 53, 55
parenteral, xi, 167, 168, 169, 170, 171, 172, 173, 174, 175
parents, 180
paresthesias, 22, 83
Paris, 50
particles, xii, 82, 169, 173, 178
pathogenesis, 16, 94, 173
pathogens, 82
pathology, 28, 95, 101
Pathophysiological, 8
pathophysiology, 20, 94, 100
pathways, 120, 163
patients, ix, x, xi, 3, 8, 9, 11, 19, 20, 23, 26, 56, 59, 73, 80, 90, 91, 93, 94, 95, 96, 97, 100, 101, 102, 103, 105, 106, 107, 109, 110, 111, 112, 113, 114, 115, 116, 122, 128, 131, 159, 160, 162, 167, 168, 170, 171, 173, 174, 179, 180
Pb, 82, 121, 131
PCBs, viii, 23, 24, 57, 62, 63, 67, 72, 77, 81, 84, 85, 91, 121, 131, 158
PCDD/Fs, 57, 121
pediatric, 171, 172, 173, 178
pediatric patients, 171, 172, 173, 178
peptide, 51
per capita, 20
perceptions, viii, 61, 63, 70, 74
periodontal, 50
periodontal disease, 50
periodontitis, 56
peripheral arterial disease, 114
peripheral blood, 115
peripheral blood mononuclear cell, 115
peripheral neuropathy, 172
peripheral vascular disease, 108
peri-urban, 86
Perna perna, 160
peroxidation, 111
pesticide, 84
pesticides, 57, 82, 120, 158, 163
pests, 82
petroleum, 83
petroleum products, 83
pH, 40
pharmaceutical, 111, 158
pharmacology, 29
pharynx, x, 12, 41, 117
phenolic, 130
phenylalanine, 38, 45
Phenylalanine, 48
Philadelphia, 105
phosphates, 163
phosphatidylcholine, 33, 57
phospholipids, 4, 10, 101, 103, 168
phosphorus, 2
photoreceptor, 172
physical activity, 129, 147, 149, 154
physical exercise, 130
physiological, 128
phytosterols, 169, 173
pilot study, 101
pituitary, 10
placebo, ix, 80, 90, 105, 106, 108, 114, 180
placenta, 17, 22, 136
planar, 35
plankton, 158
planning, 76
plant sterols, 173
plants, 40, 58, 82, 83, 85, 130
plaque, 95, 97
plaques, 96, 97, 100, 101, 123
plasma, 4, 20, 28, 34, 54, 96, 100, 101, 103, 111, 116, 121, 124, 129, 133, 154, 169, 174
plasma membrane, 101
plastic, 82
plastics, 84
platelet, 9, 12, 43, 104, 111, 112, 115, 119, 125, 130, 169
Platelet, 9, 112
platelet aggregation, 9, 12, 43, 111, 112, 115, 169
platelet count, 119
play, 2, 14, 19, 20, 22, 94, 100, 101, 127, 170, 171
pleasure, 70
pneumonia, 56
poison, 159, 164

poisoning, 23, 159, 161, 162, 164, 165
poisonous, 83, 159
Poland, 127, 133
policy makers, viii, 62
pollen, xii, 177
pollutant, 39
pollutants, 2, 18, 23, 38, 53, 82, 83, 86, 120, 163
pollution, 81, 82, 83, 87, 88
polybrominated diphenyl ethers, 22
polychlorinated biphenyls (PCBs), viii, 62, 77, 81, 121, 131
polycyclic aromatic hydrocarbon, 14, 15, 158
polyether, 162
polymorphism, 15, 31, 57, 132
Polynesia, 123, 131
Polynesian, 121
polyunsaturated fat, 2, 5, 9, 17, 19, 20, 21, 25, 27, 28, 32, 34, 38, 49, 50, 54, 75, 79, 80, 90, 96, 103, 113, 114, 115, 118, 125, 127, 131, 132, 150, 175
polyunsaturated fatty acid, 2, 5, 9, 17, 19, 20, 21, 25, 27, 28, 32, 34, 38, 49, 50, 54, 75, 79, 80, 90, 96, 103, 113, 114, 115, 118, 125, 127, 131, 132, 150, 175
polyvinyl chloride, 84
poor, 18, 19, 46, 100, 144, 168
population, vii, xi, 1, 3, 6, 7, 8, 10, 15, 16, 17, 21, 27, 28, 29, 31, 34, 35, 37, 45, 46, 50, 52, 55, 58, 65, 66, 80, 91, 95, 100, 113, 121, 123, 124, 126, 127, 132, 133, 143, 146, 149, 153, 154, 158, 165
population group, 123
pork, 85
portal hypertension, 168
positive correlation, ix, 41, 94, 96, 101
positive relation, 12
positive relationship, 12
postmenopausal, 9, 29, 30, 51
postmenopausal women, 9, 29, 30, 51
postoperative, 173
postpartum, 21, 34
postpartum depression, 21, 34
postponement, 19
potassium, 40
poultry, 54, 62, 81, 126, 129
poverty, 86
poverty alleviation, 86
powder, xii, 177
power, 83
predators, 22, 83, 85
predatory fishes, 56
predictors, 51, 94, 100, 133
preeclampsia, 49, 143
preference, 70
pregnancy, xi, 16, 17, 18, 21, 24, 25, 32, 33, 34, 53, 135, 137, 139, 140, 141, 143, 144, 146, 147, 148, 149, 150, 151, 152, 153, 154
pregnant, xi, 18, 19, 24, 25, 35, 42, 44, 52, 71, 72, 123, 135, 136, 137, 138, 139, 140, 141, 142, 143, 145, 152, 153, 154
pregnant women, xi, 18, 19, 25, 35, 71, 72, 135, 136, 137, 139, 140, 141, 142, 143, 152, 153, 154
premature infant, 168
prematurity, xi, 167, 168, 172
premenopausal, 30
prenatal care, 150
preservative, 136
pressure, 8, 9, 28, 29, 57, 88, 95, 96, 98, 99, 101, 123, 136
preterm delivery, 137
preterm infants, 174
prevention, x, xi, 2, 3, 5, 12, 26, 28, 29, 30, 38, 43, 49, 51, 52, 53, 55, 95, 103, 107, 109, 113, 118, 126, 130, 132, 157, 167
preventive, 130
prices, 163
private, 138
probability, 64
producers, 87, 158
production, 16, 17, 23, 38, 40, 44, 53, 73, 78, 82, 83, 84, 85, 86, 87, 89, 101, 102, 112, 115, 158, 163, 165, 169, 174
production costs, 158
profitability, 158
prognosis, 173
program, 34, 63, 73, 143, 145
proinflammatory, 100, 112, 115, 174
pro-inflammatory, 10, 94, 100, 112, , 128, 169, 170
proliferation, 111, 112, 127, 174
promoter, 159, 162
prostaglandin, 101, 112, 169, 170
prostaglandins, 17, 125, 173
prostate, vii, x, 1, 14, 31, 41, 48, 56, 117, 118, 120, 124, 126, 127, 132, 133
prostate cancer, 14, 31, 48, 56, 118, 126, 127, 132, 133
proteases, 102
protection, 2, 6, 10, 12, 27, 41, 71, 85, 90, 126
protein, vii, viii, xii, 2, 37, 38, 39, 45, 46, 48, 55, 57, 62, 63, 69, 70, 71, 77, 78, 94, 100, 112, 128, 129, 136, 158, 169, 170, 174, 177, 179, 180

proteins, vii, xii, 1, 22, 45, 62, 94, 100, 102, 112, 115, 131, 178, 180
proteolysis, 102
protocol, 64, 90, 171
protocols, 74, 178
psoriasis, 32
PSP, xi, 157, 159, 160, 164
psychiatric disorder, 20, 90
psychiatric disorders, 20, 90
psychiatric illness, vii, 1
public, viii, xii, 25, 62, 63, 64, 71, 73, 136, 138, 151, 164, 177
public health, xii, 25, 62, 63, 71, 73, 136, 151, 164, 177
public policy, 62
PUFA, 5, 6, 11, 20, 59, 94, 97, 98, 101, 103, 137
PUFAs, ix, 6, 24, 38, 80, 93, 94, 100, 101, 103, 125, 128, 136, 137, 146
pumps, 84
purification, 158
PVC, 82
pyrene, 163

Q

quality of life, 102
quartile, 18, 21, 43
Quebec, 140
questioning, 40
questionnaire, 64, 121, 128, 143, 145, 146, 155

R

race, 8, 139, 142, 147, 148, 149
racial groups, 142
radio, 71
rain, 40, 83
rainwater, 22
random, 107
range, 18, 42, 66, 71, 73, 86, 142
rash, 171
rats, 51, 82, 84, 101, 170, 174
reactive oxygen, 16, 83
reactive oxygen species, 83
reactivity, 9
reading, 88
reality, 173
recall, 129, 138, 142
receptors, 62, 73, 80, 83, 101, 161
recognition, 17
recovery, 56
recreation, 64, 70, 71
recreational, 63, 70, 76, 154
recreational areas, 63
rectum, x, 12, 41, 117
red blood cells, 20, 121
red meat, 13, 62, 69, 129
redox, 131
refining, 83
regeneration, 20
regional, 24, 39, 63, 71, 72
regression, x, 41, 106, 107, 109, 114, 146
regression analysis, 41
regular, 2, 10, 16, 25, 85, 110, 130, 146, 147
regulation, 20, 87, 91
regulators, 62
relapse, 52, 123
relationship, ix, x, 6, 7, 8, 11, 12, 13, 16, 17, 18, 21, 27, 28, 32, 34, 94, 106, 117, 118, 120, 122, 126, 130
relationships, 133
relevance, 55, 94, 101
remediation, 62
renal, vii, ix, 1, 15, 31, 41, 53, 59, 93, 94, 100, 101, 102, 103, 104
renal cell carcinoma, 15, 31, 59
renal disease, vii, ix, 1, 93, 94, 100, 101
renal failure, 94, 104
reperfusion, 174
reproduction, 139, 158
Research and Development, 151
residues, 158
resistance, ix, 8, 93, 102, 103, 112
resolution, 171
resource management, 88
resources, 39, 49, 81, 87
respiratory, xii, 2, 51, 57, 160, 161, 165, 177
respiratory failure, 160
responsiveness, 136
restaurants, 11
restenosis, 114
retina, 17, 172, 175
retinopathy, xi, 10, 167, 172
retinopathy of prematurity, xi, 167, 172
rheumatoid arthritis, vii, ix, 1, 2, 16, 32, 81, 90, 93
rhinitis, 178
rhodopsin, 89
rhythms, 123
rice, xii, 45, 46, 177

rickets, 55, 56, 58
risk assessment, 75
risk factors, ix, 5, 8, 12, 40, 42, 43, 46, 49, 52, 80, 89, 93, 94, 100, 115, 118, 123, 127, 130, 143, 149
risk management, 63, 71
risk perception, 154
risks, vii, ix, x, xi, 1, 2, 15, 18, 24, 25, 26, 30, 31, 34, 35, 51, 56, 61, 62, 63, 64, 67, 69, 70, 71, 72, 73, 74, 75, 76, 77, 84, 88, 90, 101, 125, 129, 130, 131, 132, 135, 136, 138, 145, 150, 153, 157, 158, 163
rivers, 82, 83, 87, 122, 158
rodent, 170
rodenticides, 82
Rome, 51, 89, 90
runoff, 83
rural, 46, 63, 86, 87
rural development, 87
rye, xii, 177

S

safety, xi, 66, 72, 78, 121, 138, 158
Salen, 29, 89
salmon, 11, 14, 22, 23, 24, 39, 48, 51, 53, 54, 57, 78, 83, 85, 89, 90, 121, 122, 131, 147
salt, xii, 17, 41, 178
salts, 121
saltwater, 63, 121
sample, 16, 126, 179
sarcoidosis, 45
satisfaction, 86, 87
saturated fat, viii, 2, 11, 77, 96
saturated fatty acids, 11, 96
SBP, 96, 99
scalp, 136
schistosomiasis, 82
schizophrenia, 19, 50, 55
Schizophrenia, 47
school, viii, xi, 65, 66, 77, 135, 143
sclerosis, 47, 52
scores, 17
seafood, 4, 8, 15, 19, 20, 21, 23, 24, 29, 38, 44, 62, 121, 131, 138, 153, 162, 163, 164
search, x, 41, 107, 117, 123, 163
search engine, x, 117, 123
searches, 123, 124
seasonal affective disorder, 19
Seattle, 135, 146
seawater, 81, 158
security, 78, 86, 163
sediments, 83, 84, 163
seed, 42, 82
seeds, xii, 39, 42, 43, 48, 177
seizures, 18
selecting, 44, 124
selenium, 2, 12, 121, 131, 158
self-report, viii, 77, 149
senescence, 40
sensitivity, ix, 9, 94, 95, 128, 133, 178
sensitization, xii, 177
sepsis, 51, 115
septicemia, 52
Serbia, 93, 95
series, 79, 144
serotonergic, 20
serotonin, 20, 21
serum, viii, ix, 4, 8, 9, 10, 18, 29, 44, 49, 50, 53, 56, 77, 93, 94, 96, 98, 100, 102, 114, 116, 119, 123, 124, 128, 133, 170, 171, 174
serum albumin, 96, 98, 102
services, iv
sesame, 43
severity, 43, 100, 159
sewage, 82, 158
sex, 22, 122
Seychelles, 18, 32, 33, 123, 137, 152, 153
SFDA, 122
Shanghai, 4, 26, 164
shape, 5
shellfish, xi, 4, 14, 15, 19, 21, 24, 26, 30, 49, 62, 79, 83, 85, 122, 136, 138, 139, 140, 143, 144, 146, 152, 155, 157, 158, 159, 160, 161, 162, 163, 164, 165
shock, 115
short supply, 45
short-term, 18, 161
short-term memory, 18, 161
shrimp, 24, 90, 122, 147
side effects, 168
sign, 79
signal transduction, 19, 112
signs, 10, 28
Singapore, 13, 30, 51
single test, 178
sites, 66, 67, 72, 83, 119
skeletal muscle, 103
skills, 18
skimming, 22
skin, 23, 44, 54, 55, 59, 85, 124, 180

sludge, 82
smokers, 5, 16, 27, 31, 126
smoking, xi, 11, 12, 15, 16, 31, 32, 129, 130, 135, 143, 147
smooth muscle, 111
smooth muscle cells, 111
social conflicts, 86
social environment, 152
socioeconomic, 74
sodium, 112
softener, 82
software, 95, 107
soil, 40
soils, 40
solar, vii, 37, 38, 39, 43, 44, 46, 52
solubility, xii, 178
sorting, 55
South Pacific, 31, 121
soy, 43, 128, 168, 169, 170, 171, 172, 175, 179
soy bean, 168, 169, 170, 171
soybean, xi, 42, 45, 127, 167, 168, 169, 171
Spain, 28, 31, 34, 50, 121, 122, 131, 154, 157, 158, 160, 162
spatial, 23
Special Supplemental Nutrition Program for Women, Infants, and Children, 142, 143
species, viii, 2, 15, 22, 23, 24, 25, 38, 62, 63, 64, 66, 67, 69, 70, 71, 72, 73, 78, 83, 86, 88, 120, 121, 137, 138, 139, 144, 146, 147, 149, 158, 159, 161, 163, 165
specific knowledge, 63, 67, 72
specificity, 138, 178
speech, 160
speed, 20
spinach, 42
sporadic, 55, 125
SRD, ix, 93, 94
stability, 158
stable angina, 11
stages, 81, 126
standard error, 65, 66
standards, 123, 151
starch, xii, 177
State Department, 76
statin, 113
statins, 50, 113
statistical analysis, 95
Statistical Analysis System (SAS), 64, 76
statistics, 89
stenosis, 9
steroid, 51, 53, 55
stiffness, 81
stock, 85, 160
stomach, x, 12, 41, 117, 119, 122, 124, 127, 133
storage, 38, 83
strategies, 49, 130, 172
streams, 83
strength, 6, 21, 88
stress, ix, 3, 12, 38, 40, 55, 58, 93, 95, 100, 102
stroke, vii, 1, 2, 3, 6, 7, 8, 9, 25, 27, 28, 62, 74
strokes, 6, 8, 123
stromal, 128
stromal cells, 128
students, 74, 145, 154
subsistence, 63, 70, 74
substances, 81, 83, 121, 130, 131, 158, 159
suffering, 21
sugar, 130
suicidal, 21
summaries, 59
sun, 30, 41, 51, 57
sunflower, 42, 43, 108
sunlight, 51, 59
superoxide, 12, 170
superoxide dismutase, 12, 170
supplements, vii, ix, 26, 37, 39, 43, 44, 46, 81, 90, 104, 105, 106, 109, 111, 113, 123, 124, 171
supply, viii, 21, 40, 41, 42, 44, 45, 50, 57, 77, 78, 81, 151
suppression, 10, 119, 170, 175
Surgeons, 173
surgery, 172, 173
surgical, 167
surveillance, 165
survival, 56, 102, 114, 115, 123, 168
survivors, 3
susceptibility, 22, 33, 57
sustainability, 28, 76
sustainable development, 86
Sweden, 28, 31, 58, 122, 125, 127, 152, 178
Switzerland, 91
sympathetic, 112
symptom, 165
symptoms, xii, 2, 21, 34, 160, 161, 162, 177, 178
synaptogenesis, 17
syndrome, 159, 165, 167, 173, 175
synthesis, 10, 58, 59, 70
systems, 22, 83, 84, 85, 86, 102, 162, 163
systolic blood pressure, 8, 9, 96, 99

T

tankers, 158
tanks, 83
tar, 83
target population, 136
targets, 82
taste, 70
taxonomic, xii, 177
TDI, 91
temperature, 163
temporal, 74, 152
territory, 138
test scores, 18, 19
tetrachlorodibenzo-p-dioxin, 23, 85
Texas, 180
Thai, 174
therapy, 50, 167, 171
thinking, 73
Thomson, 174
threat, 18
threshold, 4, 22
thresholds, 107
throat, 160, 162
thrombomodulin, 112
thrombosis, 6
thrombotic, 7
thromboxane, 112, 169, 170
thromboxanes, 173
thyroid, viii, 41, 49, 77, 127
thyroid cancer, 49
ticks, 82
tides, 163, 164
time, 3, 43, 44, 45, 64, 73, 78, 80, 83, 90, 95, 106, 111, 113, 125, 137, 143, 144, 150, 161, 171, 174
time consuming, 43
timing, 150
tin, 82, 160
tissue, viii, 6, 11, 22, 29, 77, 81, 83, 165, 172, 173
TNF, 10, 94, 95, 96, 97, 98, 100, 101, 103, 129, 174
TNF-alpha, 94, 96, 97, 98, 103, 129, 174
TNF-α, 94, 95, 96, 97, 100
tolerance, 171, 179
tomato, 126
total cholesterol, 95, 96
total energy, 40
total parenteral nutrition, 172, 173, 174
total product, 86
tourism, 71
tourist, 63
toxic, vii, 1, 11, 21, 23, 82, 85, 111, 121, 131, 144, 158, 159, 162, 165
toxic effect, 82, 144, 158, 162
toxic substances, 121, 131, 159
toxicities, 85
toxicity, 23, 83, 84, 153, 159, 160, 161, 163
toxicological, 23, 120
toxicology, 136
toxin, xi, 157, 159, 160, 162, 163, 165
toxins, xi, 120, 122, 157, 159, 160, 161, 162, 164, 165
trace elements, 2, 50, 131
trade, 78, 84, 86, 151
trade-off, 151
training, 87
transcription, 38, 80, 85, 112, 115, 174
transcription factor, 38, 85, 115
transcriptional, 112
transfer, xii, 84, 161, 177, 179, 180
transferrin, 95, 96
transformation, 165
transgenic, 102
transgenic mice, 102
transition, 12, 40, 171
transition metal, 12, 40
transition metal ions, 40
transplantation, 168, 171, 173
transport, 38, 73
trees, 40
triacylglycerols, 28
trial, 3, 25, 26, 29, 49, 50, 54, 57, 80, 89, 90, 91, 103, 107, 110, 112, 113, 114, 115, 137, 152
tribal, 75
tribes, 138
triglyceride, 8, 80, 95, 96, 98, 102, 112, 119, 168, 173, 175
triglycerides, 123, 169, 170
trout, 24, 39
Tryptophan, 48
tuberculosis, 47, 55
tumor, ix, xi, 10, 53, 94, 95, 115, 118, 128, 157, 169, 170, 174
tumor growth, 128
tumor necrosis factor, ix, 10, 94, 95, 115, 169, 170, 174
tumors, 14, 127, 160
tumour, 133, 159, 162, 174
type 1 diabetes, 59
type 2 diabetes, vii, 1, 8, 29, 46, 56, 57

U

ulcerative colitis, 2
ultraviolet, 38, 52, 57, 59
ultraviolet B, 52
umbilical cord, 154
umbilical cord blood, 154
UNCED, 87
uncertainty, 25, 153
UNEP, 89
UNESCO, 89
United Kingdom, 45
United Nations (UN), 28, 51, 87, 89, 90, 122
United Nations Development Program (UNDP), 89, 90
United States, 13, 39, 42, 43, 45, 46, 49, 53, 55, 58, 75, 84, 85, 122, 126, 132, 137, 143, 151, 153, 161, 168, 171, 172
unstable angina, 106
urban population, 52
urine, 22, 50, 154, 159
urticaria, xii, 178
US Department of Health and Human Services, 88
USDA, 45, 48, 58, 163
USEPA, 84

V

vaccine, 136
vacuum, 84
Valencia, 154, 180
validation, 131
validity, 138
values, 95, 96, 98, 99, 101, 107, 121, 139, 178
variability, 6, 11, 136
variables, 95, 99, 103
variance, 64, 96
variation, 14, 32, 52, 74, 122, 123, 125, 127, 137, 152, 153, 160
vascular cell adhesion molecule (VCAM), 10, 112
vascular disease, ix, 94, 100, 101, 108
vascular risk factors, 80
vascular wall, 104
vasoconstriction, 112
vasodilatation, 112, 115
vasodilation, 42
vector, 121
vegetable oil, 43, 115
vegetables, 82, 85, 126, 128, 129
vegetarians, 45, 54
vehicles, 83
Venezuela, 164
vessels, 83, 158
Victoria, vi, 88, 157
Vietnamese, 142
viscera, 161
viscosity, 9
vision, 136
visual memory, 23
vitamin A, viii, 44, 55, 77, 81
vitamin B1, 45, 46
vitamin B12, 45, 46
vitamin B6, viii, 77, 81
vitamin D, vii, viii, 2, 15, 37, 38, 39, 41, 43, 44, 45, 46, 47, 48, 49, 50, 51, 52, 53, 54, 55, 56, 57, 58, 59, 77, 81, 136
vitamin E, 114
vitamins, vii, viii, xii, 1, 2, 29, 77, 81, 177
VLDL, 8
vomiting, 178
voodoo, 159

W

Wales, 175
waste incineration, 23, 84
water, vii, viii, 37, 38, 46, 77, 81, 82, 83, 86, 87, 88, 91, 136, 144, 158, 160, 163
water quality, 144
water resources, 87
weathering, 82
weedy, 82
weight loss, 9, 29
weight reduction, 29
Western countries, 118
Western Europe, 52, 78
wheat, xii, 46, 177, 179, 180
wheezing, xii, 178
white blood cell count, 10
white women, 8
whole grain, 126, 129
windows, 139
winter, 38, 59
Wisconsin, 143, 145
withdrawal, 123
wives, 125, 132
Women's Health Initiative, 43
word of mouth, 67
workers, 125

World Health Organization (WHO), 12, 30, 46, 59, 85, 89, 91
World Trade Center, 54
worry, 39
writing, 40, 144

Y

yeast, 44
yellow fever, 82
yolk, 168
young women, 49, 139

Z

Zea mays, xii, 177, 180
zinc (Zn), viii, 2, 40, 77, 81, 82